Sharp Series

Chemistry for Kids
The Atomic Structure
Charges and Mass

Conceptual Learning

Lex Sharp

Fields of Code Inc.
Calgary, Alberta
www.fieldsofcode.com

Published by Fields of Code Inc.
Calgary, Alberta
Canada
www.fieldsofcode.ca

ISBN: 978 1 980 641322

FIRST EDITION

Table of Contents

What Are Things Made Of?

There is nearly nothing around us that is made of only one thing. Everything can be broken down into smaller pieces. Let's look at some real-world examples.

One such material made of multiple components is **steel**. The cutlery in your kitchen is typically made of stainless steel. To make steel, approximately 90% iron ore is mixed with Carbon in a complicated metallurgic process. Sometimes additional elements are added.

Another example is the **air** you breathe. It has:

- 78 % Nitrogen,
- 21% Oxygen,
- 0.9% Argon,
- 0.04% Carbon Dioxide,
- 0.06% Trace Gases.

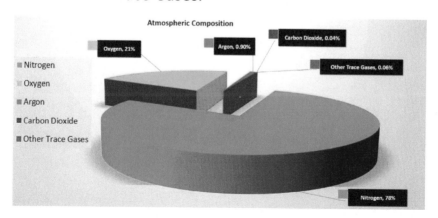

Figure 1 - Atmospheric Composition

1

Notice the atmospheric composition lists:

· Nitrogen,
· Oxygen, and
· Carbon Dioxide.

Carbon Dioxide can be separated into a Carbon and two Oxygens. However, Nitrogen and Oxygen stand alone in some unique ways.

When split further, Nitrogen and Oxygen become smaller particles. There are very few types of these particles and are often undistinguishable from one another. These inner parts are no longer chemicals and look like collections of similar gears. This is because Oxygen, and Nitrogen are instances of Elements, also known as Atoms.

An Atom or Element is of the last stage of matter before breaking down and descending into a subatomic world with no recognizable chemical "fingerprint".

Some things can be broken down into additional parts and do not obscure the original form. When gone past the atomic state an object becomes fragmented beyond recognition. Which case it may be depends on the structure of things and how far the process of splitting into smaller parts has progressed.

2

How Far Can We Zoom into Matter?

What would one see if kept zooming into matter past the larger fragments, and through the interior of the atoms?

Let's imagine for a moment matter is built similarly to Lego and investigate the parallel between the two.

The Lego pieces below construct a house. If one was to break it in two, pieces would remain recognizable as part of the whole.

In the next arrangement, Lego pieces are forming a heart. After breaking the model in two, one could probably still recognize what it was initially. Though with a less complex structure it could be a many different things, so the remaining pieces are not as obvious as before.

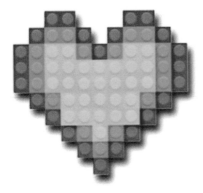

Even worse, if a collection of blocks was completely undone and left in a rubble, it's impossible to say what these made before.

Figure 2 - Lego Rubble

Matter works in a similar way. For example, a bar of Gold can be broken into two pieces but continue to be recognizable as itself. Further splits can be applied repeatedly and under a microscope would still be recognizable as Gold. Continue this process until the only parts left to split are the atoms. Breaking them leaves a disorganized pile of subatomic particles, like the Lego heap above. The disassembled fragments cannot provide clarity on the type of matter they belonged to before.

What Are Atoms Made of?

Atoms are made of smaller pieces called **Subatomic Particles** and splitting the atom separates them. Their subsets cannot be used to conclude the type of atom they came from. There is one important difference between Lego and atoms: when given the **full set** of subatomic particles from the original atom, even if all in a rubble, the atom can be accurately reconstructed.

What Is the Inside of Atoms Like?

As we dive into the comparison between Lego and subatomic particles, all that is made in the game of Lego is accomplished with blocks. Same goes for atoms and particles. Styles of bricks for dedicated areas like roofs, wheels, walls, so on, are parallel to specialized types of particles that belong in special regions of the atom and are restricted by what they can do.

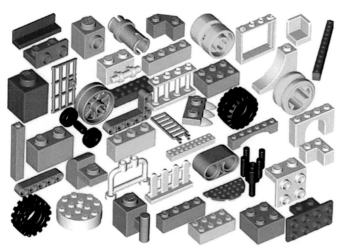

Particles fall into few categories and play their roles in the functioning of the atom. The result is a highly organized structure.

The inner cores of atoms are always made of heavier blocks while the outer shells are of a light nature and hover in cloud-like formations.

Thus, the inner world of the atom has emerged divided into two main parts: the core, and the outer shell. It roughly looks as shown.

Figure 3 - The Atomic Solar System Model

This is an approximate model of the atom that is known to be inaccurate. Nevertheless, it is useful when a simplified view of the atomic structure is needed.

The diagram is referred to as the **Atomic Solar System Model** and was suggested by scientist Niels Bohr in 1913. It is also known therefore as the **Bohr Model**.

The individual blocks, or rather particles, that make up the Bohr Model are named below. The small blue particles hovering in orbits around the larger core are called **electrons**.

Electrons reside in outer shells often referred to as clouds, where they move energetically. The electron cloud is a fundamental part of the atom. Wherever the core moves, the cloud goes with it and normally stays attached. Sometimes parts of the cloud glue on to other more poweful atoms and are consequently lost, or act as glue between a pair of atoms that stick together.

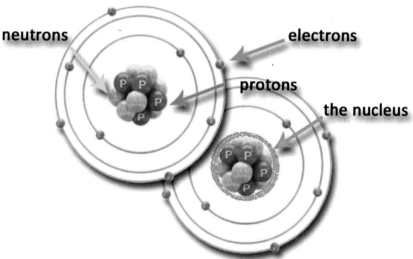

Figure 4 - Subatomic Particles

The core, or center pack is called the **nucleus**. Particles called **protons** and **neutrons** live there and cluster tightly. The nucleus exerts a strong attraction to the atom's electron cloud and keeps everything together.

7

The Bohr solar system model got its name from the resemblance of the nucleus to the sun, and that of the electrons to the planets of the solar system.

Since at the time the model was established electrons were believed to hover steadily in round or elliptical trajectories surrounding the core, the comparison prevailed.

Figure 5 - Comparison to the Solar System

The reason the solar system model is not accurate is electrons don't float around in such tidy circular orbits. They navigate in complicated symmetrical 3D paths around the nucleus.

The examples in the figure below show specific 3D spaces equivalent to two types of electron orbits. Imagine the nucleus hidden deep in the center of each image. The proper name for an orbit is orbital. Many types of orbital shapes exist, but we won't dive into this topic in the current volume.

Figure 6 - Electron Orbits

This book, as many others, continues to use the solar system model at times to help understand some important basics that do not depend on the actual shapes formed by the orbits of electrons.

Subatomic Particles

Protons and neutrons are joined together in the nucleus and do not part easily. An amazingly strong force is needed to pull them apart.

If you liked the Lego metaphor, imagine these particles to be blocks stubbornly stuck together. The force that keeps the nucleus from flying apart is known in physics and chemistry as **The Strong Force**. It has additional names, the Strong Nuclear Force, and the **Strong Interaction**. This is the strongest force we know in the universe so far.

Nucleons

Protons and neutrons are known collectively as nucleons. This is a name given to any subatomic particle that resides inside the nucleus, namely protons and neutrons.

Particle Charges

The p^+, n^0, and e^- notations will be encountered frequently for proton, neutron, and electron respectively. These show up in diagrams, chemical equations, and formulas. This book will use the notations frequently to reinforce the particles' charges.

The atom shown in the next figure has:

- 2 red spheres inside the nucleus representing the **protons**.
- 2 yellow spheres inside the nucleus representing the **neutrons**.
- 2 blue spheres hovering in orbits representing the **electrons**.

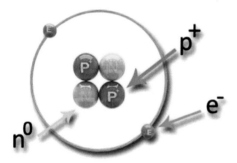

Figure 7 - Subatomic Particles and Charges

Amusingly, because of a disproportionately far distance at which electrons hover around the core, the atom is mostly made of empty space.

In theory, we can imagine an ideal elemental form of the atom to have a proton count equal to its neutron count and to that of its electrons. In other words,

the p^+ count = the e^- count = the n^0 count.

However, this is extremely rare for real life atoms. The number of neutrons is most often larger than the proton counts, and electrons get passed between atoms with relative ease during chemical reactions, leaving one atom with more and another with less electrons.

11

The **p⁺, e⁻, n⁰** shorthand reveals specific charges as part of the notation:

- electron: **e⁻**, negative charge,
- proton: **p⁺**, positive charge,
- neutron: **n⁰**, no charge, i.e. neutral.

The name "neutron" almost sounds like "neutral". Imagine the *-on* ending in the word neutron being replaced by *-al*. This is an easy way to remember it:

neutron → neutr~~on~~ + *al* → neutral.

Opposite and Identical Charges

The plus, minus, and zero superscripts represent the charge of each particle. Charges are similar to the poles of magnets, opposite charges attract, and like charges repel. Such particles will either reposition closer or further apart in each other's proximity.

Each particle type has its own designated **fixed charge** that is maintained consistently. Protons are always positive, neutrons are always neutral, and electrons are always negative.

Charge, or the lack of, is fundamental to the nature of every subatomic particle and collectively determine the inner dynamics of the atom.

Protons (**p⁺**) and electrons (**e⁻**) are two particles that are moved towards each other by their opposite charges.

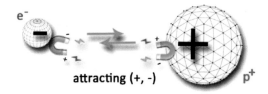

Figure 8 - Opposite Charges Attract

Alternately, a group made of only protons (**p⁺**) will repel. The same is true for a group made of only electrons (**e⁻**).

Neutrons

Neutrons are particles that have no attracting or repelling properties as far as we know today. The lack of charge is shown by a zero in the superscript of the n^0 notation. The expression *"the charge is zero"* is often used instead of saying *"there is no charge"* and has the same meaning. It can also be said the particle is neutral.

All atoms exist in **variations** called Isotopes, based on several possible neutron counts. Neutrons and their counts play an important and mysterious role in gluing the nucleus together, which is discussed a few sections further.

Electrons

Electrons are a very active part of the atom. They are responsible for chemical reactions, delivering electrical energy through wires, brain activity and a lot more.

Electrons are a part of the atom most of the time, and hover in orbitals of various shapes around the nucleus.

13

In theory electrons can exists by themselves, however due to their very "sticky" magnetic nature, they'll glue on to most things they pass by. It happens either as individual electrons, or as part of a larger group like a "sea of electrons" in plasma where electrons exist side by side with other atomic fragments (ions), but never quite alone as free-floating particles on their own.

When attached to the atom, electrons never come too close together. They experience strong repulsion to each other due to identical negative charges that repel similarly to like-pole magnets. The result is electrons will hover as far apart as possible but remain tied to the parent atom at the same time.

Repulsion causes tension between the electrons as they are trapped in various orbital levels. Despite this, electrons are in perpetual motion and fidget with great energy and cannot be completely bound. It is surprising they manage to stay together despite forces that propel them continuously apart.

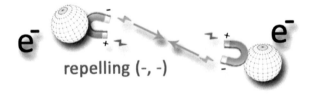

repelling (-, -)

Figure 9 - Negative Charges Repel

The force responsible for repulsion is called the **Electromagnetic Force**. It has the same effect as

14

magnetism with like-poles of magnets. This raises the question: what keeps the electrons attached to the nucleus? With so much repulsion how is it that they do not push each other out of the way and far from the influence of the atom?

The answer is electron repulsions (blue arrows below) are not the only forces acting on the electron cloud. The positive proton core is attracting the negatively charged electrons (red arrows below) and ropes in the cloud.

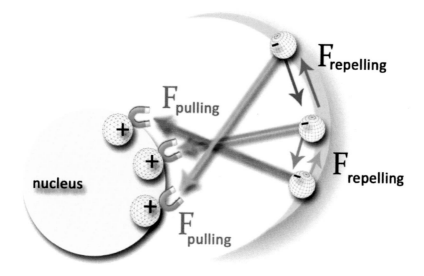

Figure 10 - Equilibrium and Electrons

The two types of forces act in opposition to each other. The result is the electrons neither fly apart, nor get too close to collapse into the nucleus. This system maintains positional **equilibrium**.

15

Equilibrium

Equilibrium is a state in which there are many competing forces on an object. The participating forces are continuously active. During equilibrium all forces settle in a way that looks like nothing is happening at all. It can be thought of as no force is winning. When comparing forces to the children in a tug of war game, equilibrium would be equivalent to a stalemate.

Figure 11 - Equilibrium, a Tug of War Stalemate

Equilibrium can also be imagined as the base state of a spring. One can stretch the two ends away from each other but springs tend to return to their original size when let go. Once restored to its original state, the spring is in **equilibrium**. The forces acting on the spring upon release are at war with each other pushing and pulling until eventually they stabilize into a stalemate and finally movement stops.

The forces acting on a system while transitioning to equilibrium are called Restoring Forces.

Facts about Electromagnetism

Here are a few interesting facts about the Electromagnetic Force. It is one of the four fundamental forces known in physics:

(i)   the Strong Force,
(ii)   the Weak Force,
(iii)   the Electromagnetic Force,
(iv)   Gravity.

The Strong and Weak forces only work in the vicinity of the nucleus, i.e. have an effect over limited distances. Beyond certain distances from the nucleus, particles are no longer affected at all.

Gravity and the Electromagnetic force on the other hand have an infinite range of action. However, that is not to say their effects stay the same everywhere.

The intensity of Gravity and Electromagnetism is reduced with distance by a mathematical formula called the Inverse Square Law. The law describes how forces spread outwards from the origin in a cone-like formation. This is easily imagined as straws spreading out of a glass, close together at the bottom and spread apart at the top. The longer the straw, the wider the spread becomes.

The next figure shows the cone-like formation mentioned above. The field lines, similarly to straws, travel through widening divisions of space called sections. Electromagnetic forces are most concentrated and thus strongest at the origin. The origin is the tip of the cone. The influence of electromagnetism on the environment depends on the collection of forces working together over the area of a section.

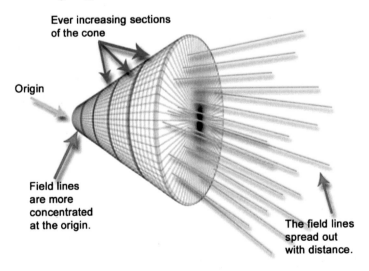

Figure 12 - Inverse Square Law

The forces, or electromagnetic field lines travel through ever widening sections of the cone. Field lines spread out evenly over the entire obtainable area of each section.

A cone section pierced by a high density of forces has a strong electromagnetic effect. When the opposite happens, low density piercing generates a low electromagnetic effect.

Thus, far from the origin the effect of the electromagnetic force collective becomes diluted into infinitely broader circles and is increasingly fading. Nevertheless, its impact can never be said to be totally zero, and thus has an infinite range.

Protons

Protons, like neutrons, belong to a class of particles called **nucleons**, and reside inside the nucleus.

Just like electrons, protons have like-charges. In this case the charges are positive. Protons experience strong repulsion and here also occurs due to the **Electromagnetic Force**. The resulting behavior is once again comparable to the force field experienced by a magnet's like-poles.

Earlier we explored why the electron cloud stays together in spite of persistent repulsion. The same question can be asked about protons, why does the electromagnetic repulsion of positive charges not catapult all protons apart?

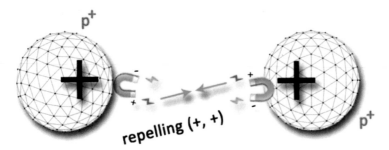

Figure 13 - Positive Charges Repel

The force that keeps protons together inside the nucleus is the **Strong Force**. Without it the nucleus would gust its particles in all directions. But that is not the complete story. The Strong Force can only work its unexplained magic if the nucleus is mixed up with neutrons.

Figure 14 - Neutrons and Shielding

20

A nucleus made of only protons will not stay together. At least one neutron[1] is needed for every proton to maintain cohesion, except for the nuclei of Hydrogen and Helium. Neutrons stand between protons and prevent the nucleus from bursting apart in a process called **Nuclear Fission**.

Neutrons create walls of shielding that limit the effects of the positive repulsion. However, the geometry of **large atoms** is such that a one to one ratio of protons to neutrons is not sufficient to maintain the stability of its protective shielding. The larger the proton count, the more neutrons must be present, or shielding will fail.

Even the parity of the proton count, i.e. the atomic number, makes a difference. Elements with an even[2] proton count accomplish stability with more ease than those with odd counts.

Contrary to intuition, a larger than necessary collection of neutrons will also cause instabilities through an imbalance of energy within the nucleus. Just as neutrons stabilize protons, the protons counteract the instability of neutrons via their energy of repulsion.

---

[1] Chemistry: The Molecular Science, By John W. Moore, Conrad L. Stanitski, page 789, Nelson Education, Jan. 24, 2014.
[2] Chemistry: The Molecular Science, By John W. Moore, Conrad L. Stanitski, page 791, Nelson Education, Jan. 24, 2014.

By themselves, neutrons are very unstable[3], with an average lifetime of less than 30 minutes. When there aren't enough protons to create equilibrium, the extra neutrons spontaneously decay and fracture the atom into smaller pieces that may or may not be stable as well. Typically, this applies to elements that are on the heavier side of the periodic table, with a few exceptions.

[3] https://journals.aps.org/pr/abstract/10.1103/PhysRev.83.349

Atomic Mass

The idea of subatomic particles having mass, or not, is one of the most fascinating realities of Particle Physics. Particles either have distinct mass or are mostly or entirely made of energy. This fact in itself is not so riveting until we remember that normally energy flows, leaks, gets mixed with other energies. One such example is heat transfer. Then we can truly marvel at energy-particles and how they persist in their own bubbles without diffusing into an ambient background filled with a multitude of free-flowing energy types.

The three major subatomic particles have one important difference between them: **mass**. Both protons and neutrons have dramatically more mass than electrons. It is also known that neutrons are slightly heavier than protons.

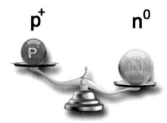

Figure 15 - Neutrons Are Heavier Than Protons

Electrons are made mostly of energy, though they too have a bit of mass but extremely small by comparison. Electron mass is considered negligible in relation to the total mass of the atom and is thus most often disregarded.

Nevertheless, the precise details can sometimes be important.

In the following model, both the p^+ and the n^0, are replaced by their more generic average, the nucleon. The nucleon mass approximately equals the joint mass of 2,000 electrons.

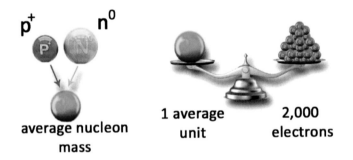

$$p^+ \qquad n^0$$

average nucleon mass	1 average unit 2,000 electrons

Figure 16 - Electron Mass

Nucleons are considered to be the only relevant particles to the total mass of the atom, formally known as the **Atomic Mass**.

The following relation applies to every element **variation**:

Atomic Mass = **total nucleons** (measured in atomic mass units).

Notice the numerical value above is a particle count. However, to designate mass properly as indicated in the term name, each nucleon in the count is perceived as a contributor of a consistent average mass of a p^+ and n^0, as represented in the previous figure. Thus,

$$1 \text{ nucleon} \longleftrightarrow 1 \text{ atomic mass unit.}$$

This is how the nucleon became the standard for **one atomic mass unit**, in short **amu** or **AMU**. The amu has been adopted as a formal unit. There is no need for unit conversions between particles and mass units, they're the same:

$$x \text{ nucleons} \longleftrightarrow x \text{ amu.}$$

Atomic Mass is also known as

- **Atomic Mass Number,**
- **Mass Number,**
- **Symbol A**, and
- **Nucleon Number.**

Read more about AMU in the *Atomic Mass Units* section.

It should be easy to see now that Atomic Mass is an attribute of a specific element variation and should not be confused with the average mass of all variations.

Average mass is described in the definition of the Atomic Weight covered in a later chapter.

Average and Atomic Mass

The average related to the Atomic Mass is shown in *Figure 17 - Average Nucleons and Mass*, between

25

protons and neutrons. Fortunately, the difference of mass between them is approximately zero ($\cong$)[4].

One way to emphasize the idea of difference in science is to use the Greek letter Delta: Δ.

Getting used to this new notation will prove useful in the future. Typically, Δ replaces the wording "*the difference between two values...*".

Thus,

$$\Delta \text{ mass} = (\text{mass of 1 } n^0) - (\text{mass of 1 } p^+),$$

$$\Delta \text{ mass} \cong 0.$$

Averaging the mass of protons and neutron into one single type of unit (the nucleon) works well because the mass values are so close. Then protons and neutrons can be dropped in favor of the more generic particle type, the nucleon, to simplify matters.

In conclusion,

$$\text{average of one nucleon mass} = \frac{1\ p^+\text{mass} + 1\ n^0 \text{ mass}}{2} \qquad \textit{(Equation 1)}$$

and,

$$\text{average nucleon mass} \cong \text{mass of 1 } p^+,$$

$$\text{average nucleon mass} \cong \text{mass of 1 } n^0.$$

---

[4] $\cong$, $\approx$, and $\simeq$ are mathematical symbols for approximate.

The next figure shows how to visualize the total Atomic Mass of an element variation as particle averages, therefore instead of

$$3p^+ + 3n^0,$$

a set of 6 simplified nucleons is used. Whether particles are averaged (on the right) or not (on the left) both cases end up with a near identical total mass.

Figure 17 - Average Nucleons and Mass

Example

Examine the following diagram.

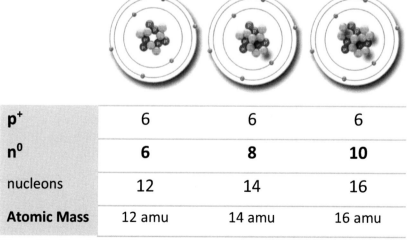

p$^+$	6	6	6
n^0	6	8	10
nucleons	12	14	16
Atomic Mass	12 amu	14 amu	16 amu

Figure 18 - Atomic Mass Examples

The diagram shows three variations of the same atom.

The Atomic Mass of each variant was calculated based on total nucleon counts.

When converting each nucleon into one mass unit, the count is used to communicate the corresponding mass. Thus, 12 nucleons become 12 atomic mass units, 14 nucleons become 14 amu, and so on.

Read more about *Mass* and *Mass Weight Comparison* in the sections that follow.

Organizing Elements

An atom (a.k.a. element) is the smallest part of matter that is still recognizable as its unique self. A few examples are: Hydrogen, Oxygen, Carbon, Gold, Silver, Iron, and there are many more.

Scientists confirm (as of 2018) the existence of a total of 118 elements so far, as naturally occurring or as man-made in a laboratory.

Chemists keep track of all known elements in a famous table called the **Periodic Table**.

The Periodic Table starts with **H** (Hydrogen), followed by **He** (Helium), both in the first row, and continues sorting elements in the order shown below.

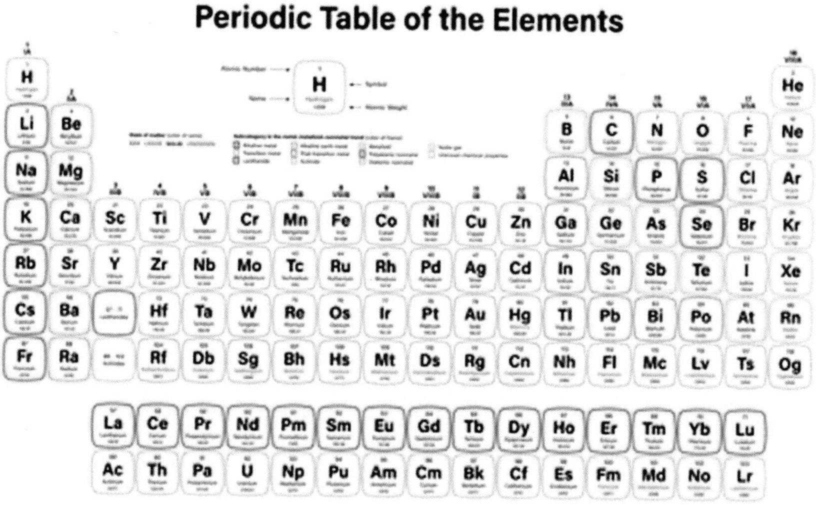

Periodic Table of the Elements

Figure 19 - Periodic Table

The sequence in which elements are listed is significant. Its earliest version was invented by Russian scientist **Dmitri Mendeleev** in the late 1800s. It is considered to be the most ingenious arrangement and its base idea is still used today. The position of elements in the periodic table was used to predict the existence and often some common properties of newly discovered elements or those assumed to be likely to exist.

Zoom into the table cells and find how each host its own unique element. There are no duplicates. Every cell has:

- a one, or two lettered symbol that is an abbreviation of the atom's Latin name,
- a **whole number** used as an ID,
- a **decimal number** representing an average.

Periodic tables are often presented in artistic styles. Some styles show the full English name of the element. The following are a couple of examples.

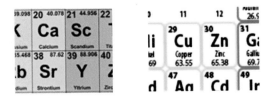

Figure 20 - Periodic Table Styles

Sorting in Rows and Columns

Sometimes, the atom is analyzed as if it was a sphere. The nucleus resides in the center containing the protons and neutrons. Layers of electron clouds surround the nucleus, each hovering at a specific distance from the middle. The outer layer gives the sphere its boundary and is where the **atomic radius** is measured.

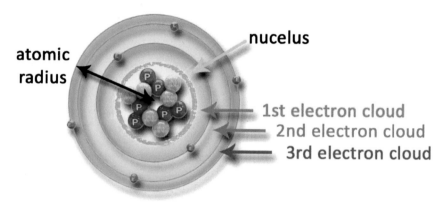

Figure 21 - Electron Clouds

Interestingly, atomic radii are not always proportional to the number of subatomic particles contained.

Atomic radii trends are most noticeable in the rows and columns of the periodic table.

Columns

When moving from cell to cell within a column, atoms are observed to have increasingly larger radii, with the largest at the bottom.

This makes sense since with each cell's nucleon count become significantly larger. For instance, the first column of the periodic table is shown next. It displays the average nucleon count of all assumed variations for

each element. Notice the resulting mass grows
exponentially from the top downwards.

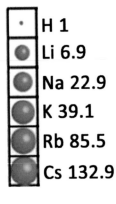

H 1

Li 6.9

Na 22.9

K 39.1

Rb 85.5

Cs 132.9

The number of electrons
accompanying such large nucleii is
also high. The extra electrons
cannot be all fitted into the lower
orbits (i.e. closer to the core).
Instead, the electrons get
distributed among new electron
clouds in higher orbits, thus forcing
the atomic radius to stretch.

Figure 22 - Atomic Radii in Columns

Rows

In turn, examining atoms across rows proves that atomic
radii are shrinking in some areas, and are inconsistent
elsewhere. This is despite nucleon counts being reliably
larger from left to right in every row.

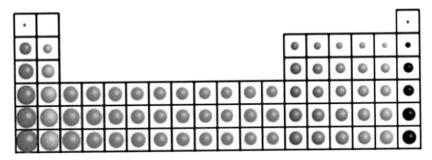

Figure 23 – Atomic Radii

The phenomenon can be explained by the number of
electrons. Most atoms residing within the same row

have an identical number of layers that sustain electron clouds. There are a few exceptions, but for most, the clouds have plenty of space. All the electrons can be contained without a need to create additional orbits.

The last cloud (the boundary) hovers in the same orbit for many of the atoms in that row. These atoms maintain identical radii regardless of particle counts.

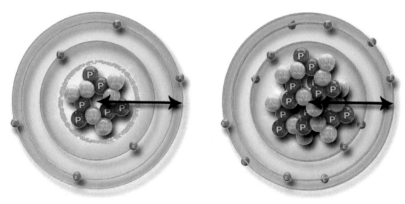

Figure 24 - Identical Atomic Radii

The positive charge of the nucleus increases in atoms from left to right in every row. This is because the number of protons contributing the charge increases by one from cell to cell. Electron counts also increase one by one from left to right.

While tracking these counts we discover that every once in a while an atom has some lucky number of protons that is able to overcome the outermost electron cloud and establish a much stronger electromagnetic pull that tightens the "atomic belt" (see the atom on the right in the figure below). The pull does not change the number of orbits, but it does drag them closer inwards creating a more crowded environment and a shorter atomic radius. This doesn't happen for all elements consistently, so there are several inconsistent trends in each row.

Figure 25 - Shrinking Atomic Radii

In conclusion, the sum of protons and neutrons determines the atom's mass, weight, nucleus size, but not necessarily its radius unless comparing atoms in the same column.

Telling Atoms Apart

Scientists in the late 1800s, and early nineteen-hundreds have speculated much about the strength of positive charges identified with atoms. Today we know this charge originates from the protons contained in the core.

At the time it was not well understood how subatomic particles are organized in and around the nucleus. Most experiment ideas of the time revolved around the atom's positive and negative charges.

Researchers managed to measure the intensity of the positive charge emanating from several atoms. They wanted to know how it relates to the logical position of elements in Mendeleev's suggested order for the periodic table. However, since the original periodic table misplaced several of the elements, the mistake obscured conclusions about the true nature of protons.

A scientist by the name **Henry Moseley** designed an experiment in 1913 that revealed with clarity that the **identity** of each atom is connected to the number of its **proton count**. Or as it was perceived then, the units measuring an atom's positive charge.

The order of elements in the periodic table was adjusted over time in light of this discovery. Moseley's experiment revealed that the difference between two consecutive atoms was exactly **one unit of positive charge**, i.e. as we know it today: one proton, and that such units are never fractioned. This is a way of saying the number of protons was found to always be present in whole numbers.

The expression "positive charge" is still used in books and articles today by some scientists as a substitute for the "proton count" wording.

Atomic Identity

If one could observe the following two atoms under an electron-microscope, it would be interesting to know what type of elements they are. Are these Hydrogen, Iron, or maybe Oxygen? Is it possible the two represent the same atom?

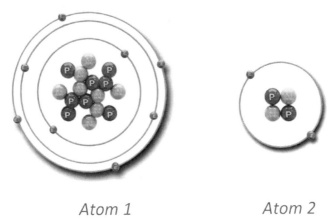

Atom 1　　　　　*Atom 2*

Figure 26 - Comparing Atoms

The answer is the two diagrams show two very different type of atoms. They are:

Atom 1 = Oxygen, with a count of 8 p⁺
Atom 2 = Helium, with a count of 2 p⁺

The only sure way to know these two are different atoms, is to verify they have a different proton count.

Typical Periodic Table Cells

The typical periodic table cell has the following arrangement:

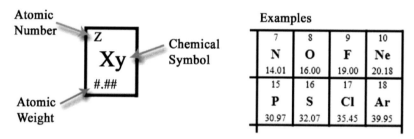

Figure 27 - Typical Periodic Table Cell

Atomic Number

We know today based on Moseley's work the **proton count** is the atom's unique and definite **ID** into the periodic table. *Figure 26 - Comparing Atoms* shows proton counts (the red spheres) of 2 and 8. The corresponding periodic table cells are found in the first two rows and show the whole numbers 2 and 8. From there the atoms are easily identified as **Helium** and **Oxygen** respectively based on their chemical symbols.

Periodic Table

					18 8A
13 3A	**14** 4A	**15** 5A	**16** 6A	**17** 7A	**2** **He** 4.003
5 **B** 10.81	6 **C** 12.01	7 **N** 14.01	8 **O** 16.00	9 **F** 19.00	10 **Ne** 20.18
13 **Al** 26.98	14 **Si** 28.09	15 **P** 30.97	16 **S** 32.07	17 **Cl** 35.45	18 **Ar** 39.95

8	9	10	11	12						

26	27	28	29	30	31	32	33	34	35	36

Figure 28 - He and O in the Periodic Table

After Mosely's experiments, the proton count was formally named the **Atomic Number**.

The **Atomic Number** is also known as

- Proton Number,
- Symbol Z ,

and identifies the atom **uniquely**.

3.69	63.55	65.38	69.72	72.5
46	47	48	49	50
'd	**Ag**	**Cd**	**In**	**Sr**
16.4	107.9	12	114.8	118
78	79	0	81	82
?t	**Au**	**Hg**	**Tl**	**Pł**
5.1	197.0	200.6	204.4	207

Figure 29 - Atomic Number

A proton cannot be fractioned further into lesser pieces such as half, quarter, etc., without destroying the atom. As a result, the **Atomic Number** must always be a **whole number**, so no decimals can exist for this concept.

38

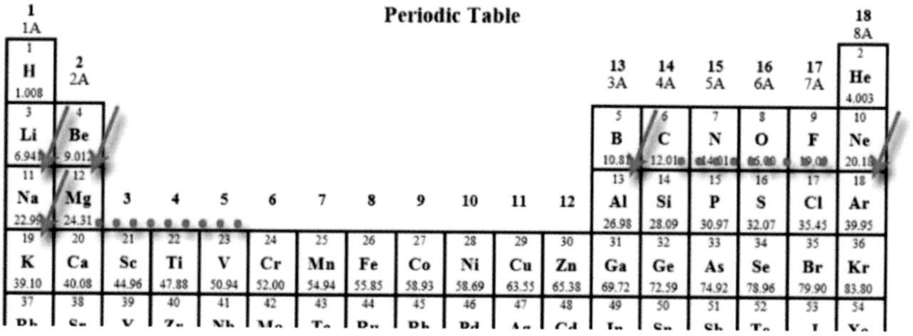

Figure 30 - Periodic Table Cell IDs

Atomic Numbers increase by one from cell to cell and left to right in the periodic table. After the end of each row, this process continues in the left most position of the subsequent row.

All types of atomic variations share the same proton count and thus an identical Atomic Number within the variation group, which clearly gets them assigned the same periodic table cell. Accordingly, it is not the isotopes that get a placement in the periodic table but rather the parent atom they belong to.

An atom experiencing a drop or increase in proton count becomes a different element than it was before. This can only happen as part of

- Nuclear Fission, p^+ count drops, or
- Nuclear Fusion, p^+ count increases.

An example of nuclear fusion is the compressing of Hydrogen atoms into Helium which fuels the Sun and other stars. As for fission, it is the process in which

39

elements break down into lighter atoms by fracturing the nucleus. Elements that are actively experiencing fission are called radioactive. There are many atoms that are naturally radioactive on Earth, others occur from man-made technologies such as nuclear power plants and nuclear weapons.

Chemical Symbol and Name

The periodic table assigns unique names and chemical symbols for every element. There used to be two styles of element designations: named and unnamed before 2016. Currently all known atoms appear in the periodic table, i.e. up to ID 118, and are all named.

.69	63.55	65.38	69.72	72.5
46	47	48	49	50
'd	Ag		ın	Sı
46.4	107.9	112.4	114.8	118
78	79	80	81	82
't	Au	Hg	Tl	Pl
45.1	197.0	200.6	204.4	207

Figure 31 - Element Symbol

All **named atoms**, have symbols made of either one or two letters.

Unnamed atoms were initially unknown but theorized to exist. At the time these were not confirmed yet. Nevertheless, they were tentatively logged in the periodic table using a three letters symbol derived from

40

their expected Atomic Number ID. Initials of the Latin numbers were used. Below one such example is described.

Unnamed Atom Example

Prior to 1994 element numbered 111 was unknown. The Latin word for the number 1 is: **Un**. The name designated for this element was: **Unununium** and was based on the following layout.

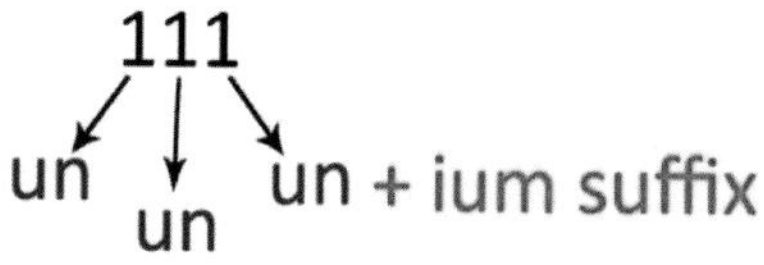

Figure 32 - Naming Unknown Elements

The *"ium"* suffix was added for consistency with other named elements. And thus, a new three letter symbol **Uuu** was attached to its cell in the periodic table.

Ru	Rh	Pd	Ag	Cı
101.1	102.9	106.4	107.9	112
76	77	78	79	8(
Os	Ir	Pt	Au	H
190.2	192.2	195.1	197.0	200
108	109	110	111	
Hs	Mt	Uun	Uuu	
(265)	(266)	(269)	(272)	

Figure 33 - Unnamed Elements

Uuu has been artificially synthesized in 1994 and was renamed as **Roentgenium**.

Legacy Names

The first original atom names were in Latin and are still used today. Many of the English names are identical to the Latin names, but some differ. A few such instances are shown next.

Symbol	Latin Name	English Name
Na	Natrium	Sodium
K	Kalium	Potassium
Pb	Plumbum	Lead
Au	Aurum	Gold

Examples of elements with identical Latin and English names are displayed below.

Symbol	Latin Name	English Name
Mg	Magnesium	Magnesium
Ca	Calcium	Calcium
Se	Selenium	Selenium
O	Oxygen	Oxygen

Older versions of the periodic tables do not reveal the atom's full name, only its symbol. You are expected to memorize the symbol and the English name for every single element.

A while back only paper versions of the periodic table were used. Today many apps are available for use with digital versions of the Periodic Table which makes the exploration more interesting.

One such is the free **Asparion Periodic Table**, you can find it at:

https://www.asparion.de/en/apps/periodictable.html.

This app displays useful properties like the element's English name, its boiling and melting points, electronegativity data, hazard symbols (pictograms), and more.

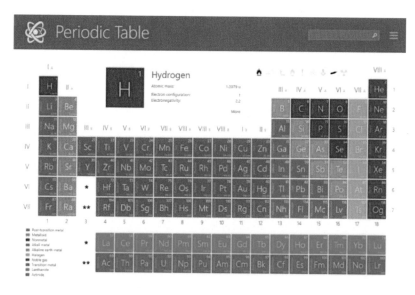

Figure 34 - Asparion Digital Periodic Table

Subatomic Particle Counts

Identical Counts

The three atoms in the next figure show p^+, n^0, and e^- counts. The specific atoms shown in fact have identical counts in real life.

$$p^+ \text{ count} = n^0 \text{ count} = e^- \text{ count}$$

This is very neat, but equalities such as this are a theoretical ideal that rarely happens in nature. When it does, we only see it in the upper rows of the periodic table.

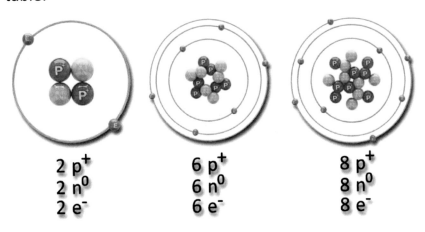

$$2\ p^+$$
$$2\ n^0$$
$$2\ e^-$$

$$6\ p^+$$
$$6\ n^0$$
$$6\ e^-$$

$$8\ p^+$$
$$8\ n^0$$
$$8\ e^-$$

Figure 35 - Atoms with Identical Particle Counts

These lucky three atoms shown seem to accomplish an ideal count most of the time. It must be said *"most of the time"* because elements have numerous neutron (n^0) variations that kick the particle count out of the norm.

The two atoms on the right (above) are significant in **mass calculation standards**. They have an even number of protons and neutrons, which generate convenient mass averages. Mathematically speaking these chosen few generate the least error and an even spread between the larger neutron mass and its proton counterpart (see *Figure 15 - Neutrons Are Heavier Than Protons*).

A quick visual scan of the periodic table lands on the listed **Atomic Numbers**, here in blue: 2, 6, and 8.

1 1A			Periodic Table														18 8A
1 **H** 1.008	2 2A											13 3A	14 4A	15 5A	16 6A	17 7A	2 **He** 4.003
3 **Li** 6.941	4 **Be** 9.012											5 **B** 10.81	6 **C** 12.01	7 **N** 14.01	8 **O** 16.00	9 **F** 19.00	10 **Ne** 20.18
11 **Na** 22.99	12 **Mg** 24.31	3	4	5	6	7	8	9	10	11	12	13 **Al** 26.98	14 **Si** 28.09	15 **P** 30.97	16 **S** 32.07	17 **Cl** 35.45	18 **Ar** 39.95

The IDs correspond to the following element names:

ID = proton count = Atomic Number	Symbol	Element Name
2 p+	He	Helium
6 p+	C	Carbon
8 p+	O	Oxygen

Another atom that should be mentioned with this batch is Silicon (Si). Its Atomic Number is 14, and its typical overall nucleon count is 28, which is also spread between protons and neutrons evenly.

45

Different Counts

When nucleon counts are not evenly spread between protons and neutrons, mass averages tend to be somewhat skewed. It will lean too much on either the protons mass or alternately on more of the neutrons, producing errors and a less reliable average. Such variations are not the best candidates for standards of average mass calculations.

Using the previous example, imagine the particle counts no longer to be equal.

He	C	O
$1\ n^0$ removed	$1\ n^0$ added $1\ e^-$ removed	$2\ n^0$ added $2\ e^-$ added

Now,

$$p^+ \text{ count} \neq n^0 \text{ count} \neq e^- \text{ count.}$$

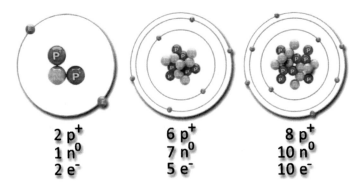

$2\ p^+$ $1\ n^0$ $2\ e^-$	$6\ p^+$ $7\ n^0$ $5\ e^-$	$8\ p^+$ $10\ n^0$ $10\ e^-$

Figure 36 - Atoms with Different Particle Counts

Notice the number of **protons** stays the same, so the related base atoms have not changed. Such differences are possible because atoms can form variations on both neutrons and electrons.

There are two possible kinds of atomic variations: **Ions** and **Isotopes**. To imagine how Ions and Isotopes belong to their "parent" atoms, elements are compared to colors and their variations to shades.

For instance, the colors Red and Blue are used as stand-ins for two separate elements, and the shades are like their Ions and Isotopes.

Like a group of shades belongs to one color, so does a number of variations rest on the same **number of protons**, i.e. the same element, regardless of how many other subatomic particles are contained.

Isotopes

Isotopes are specific variations of the **neutron** counts. The number of neutrons can either be equal, or different than the number of protons. Thus, every instance of an atom is some sort of isotope and is tagged by a hyphenated **nucleon count**. The neutrons affect the total of nucleons, and the Proton Number is always fixed.

$$\text{the nucleon count} = p^+ + n^0, \text{ thus also}$$

(Equation 2)

$$\text{Atomic Mass} = p^+ + n^0$$

and both the following conditions are acceptable:

$$p^+ \text{ count} = n^0 \text{ count, and}$$

$$p^+ \text{ count} \neq n^0 \text{ count.}$$

Isotopes are a means of grouping instances of an element based on nucleon categories and are represented by the atom's total mass.

Example

Most Boron atoms are either tagged **Boron-10**, or **Boron-11**. All others are man-made instances of Boron created in the lab with less than a second life span and have received mostly tags between Boron-6 to Boron-9, and Boron-12 to Boron-19.

Boron-10 is the only variant that has: p^+ count = n^0 count, and it too is an isotope, and is known to be stable.

Prevalence

For every isotope discovered, a **prevalence designation** is suggested in addition to its name tag. Prevalence[5] is expressed as the **percent** of atomic instances with a

---

[5] https://chemistry.sciences.ncsu.edu/msf/pdf/IsotopicMass_NaturalAbundance.pdf

given neutron count among all possible variants of the element.

For example, the two isotopes **Carbon-12** and **Carbon-13** make 98.93% and 1.07% respectively, of the total number of carbon types assumed to exist on the planet.

Ions

Ions are defined by their **electron differences**. More precisely, to be an ion the atom must have an **electron count** that is different than the proton count:

$$p^+ \text{count} \neq e^- \text{count}.$$

When electron counts are the same as the Atomic Number, the atom is not considered to be an ion. Thus, not every instance of an atom is an ion.

One More Thing...

Instances of an element can be both Isotopes and Ions at the same time[6]. For example, **Helium-4** is the most common isotope of Helium. When it loses 2 electrons it becomes an ion as well. This instance of Helium is both an Ion and an Isotope.

Atoms become more chemically reactive when electrons are lost or gained after a conversion to ion occurred. Isotopes in turn are divided into two types: stable and

---

[6] http://writing-guidelines.web.cern.ch/entries/ions-and-isotopes

radioactive. The difference in neutrons has no impact on the atom's chemical reactivity in a traditional sense.

Isotope Notation

Isotopes are important in Nuclear Chemistry, but also in other fields such as organic chemistry and **Mass Spectrometry**. A formal notation of isotopes is necessary to distinguish between them.

Isotopes are based on neutron counts, which can be either good or bad for stability. They play a role in shielding protons from each other in the core, but when overrepresented they have an opposite destabilizing effect. Therefore, isotopes are regarded not only for their various mass but also for their **stability**.

Information about the stability of an isotope is vital for deciding the safety measures required to handle such materials.

There are two main isotope notations.

The first notation uses the element's full name followed by the hyphenated nucleon count. For example,

> **Carbon-12,**
> **Helium-4**.

The second notation relies on the element's chemical symbol instead of its name and is preceded by a numerical superscript on the left above the symbol.

$$^{A}_{Z}X$$ A is the **Atomic Mass** or **nucleon count** unique to the given isotope, and as such a whole number; Z is the p^+ count, and **X** is the atom's chemical symbol.

The more frequently used simplified notation is similar:

$$^{A}X$$ Z is dropped, because the p^+ count is implied by the chemical symbol **X** which clearly identifies the given element.

In both cases the **Atomic Mass A** appears at the top left corner of the chemical symbol and leads the isotope's naming scheme. A few examples are:

$$^{12}_{6}C, \quad ^{3}_{2}He, \quad ^{11}_{5}B.$$

The extended isotope notation with both **A** and **Z** displayed, provides a quick visual cue about the proton to neutron ratio in the nucleus. Furthermore, **A** being the nucleon count, **Z** the protons, the neutron count is easily calculated by: n^0 count $= A - Z$.

The following relationships become evident:

Carbon-12: $p^+ = n^0$,

Helium-3: $p^+ > n^0$,

Boron-11 $n^0 > p^+$.

Caution...

The isotope notation should not be confused with the formatting of the **periodic table cell** (see figure below). The two list the nucleon and proton totals in reverse order.

Cell format   ·   Z is shown at the **top** of the cell,
·   the nucleon count is listed at the bottom as an average and is a decimal number.

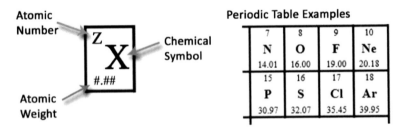

Figure 37 - Periodic Table Cell Arrangement

Isotope   ·   Z listed at the **bottom**, if at all.
notation   ·   the nucleon count A (Atomic Number) is listed at the top, and belongs to a particular isotope, rather than an average, and is a whole number.

$$_{Z}^{A}X, \text{ or } {}^{A}X.$$

Figure 38 - Isotope Notation

Ion Notation

The Ion concept is based on a previously described theoretical ideal in which

$$p^+ \text{ count} = e^- \text{ count}.$$

All atoms in their elemental, non-ionic state obey this rule. We can also say that the positive and negative charges balance each other out to describe such equalities.

An atom is **balanced** when each of its protons is coupled uniquely with an electron forming a neutral (+, -) pair.

For every proton there must be an electron counterpart and vice versa, for the atom to be neutral. Then the element is not able to form an overall atomic charge.

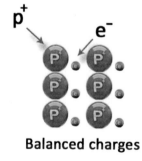

Balanced charges

Figure 39 - Balanced Ionic Charges

An ion forms when this rule is broken by either adding or removing electrons. So clearly ions are all about the electrons. An ion notation is used to show how far an imbalance has progressed. The notation does not provide information about the current number of particles, but rather about the number of unpaired charges (i.e. electrons or protons) and their type: + or -.

Ion notation consists of

- the chemical symbol, here X was used generically,
- the amount of charge c that equals the particle count causing the imbalance, and
- the type of charge: plus, or minus.

$$X^{c+} \quad \text{or} \quad X^{c-}$$

When c = 1, c can be absent.

Example Ions: Na^+, Cl^-, Cu^{2+}, Al^{3+}, O^{2-}, S^{2-}.

Notice the difference between ion and isotope notation. In an ion the sign of the charge designation always accompanies the number and appears on the right. The isotope has no sign, and the number appears on the left.

Cations: The Loss of Electrons

The loss of one or more electrons makes the atom overly **positive** and is known as a *positive Ion* or *Cation*.

This happens because of extra positive charges from unpaired protons. All the matched pairs stay neutral and the imbalance originates from whatever is left over. The figure below shows a charge of **2+** caused by 2 unpaired protons.

54

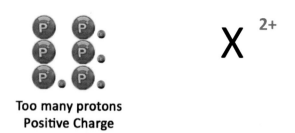

Too many protons
Positive Charge

Figure 40 - Positive Ionic Charges

Anions: The Gain of Electrons

The gain of one or more electrons makes the atom overly **negative**. The atom is then known as a **negative Ion** or **Anion**.

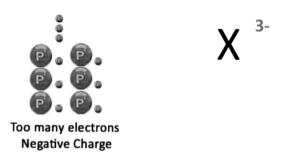

Too many electrons
Negative Charge

Figure 41 - Negative Ionic Charges

This happens when there is no equivalent number of protons to counteract the newly acquired electrons. Even though all matched pairs stay neutral, negative charges will occur from all electrons that were left uncoupled.

The figure above shows a charge of **3-** generated by 3 extra unmatched electrons.

55

Atomic Weight

The typical periodic table cell displays two numbers

- a decimal (#.##) at the bottom, and
- a whole number (Z) at the top.

.69	63.55	65.38	69.72	72.5
16	47	48	49	50
'd	**Ag**	**Cd**	**In**	**Sr**
16.4	107.9	112.4	114.8	118
78	79		81	82
't	**Au**	**Hg**	**Tl**	**Pl**
15.1	197.0	200.6	204.4	207

Figure 42 - Atomic Weight

The decimal number is called the **Atomic Weight.** It equals the **weighted average of nucleon counts** from all existing isotopes of an element. The average is "**weighted**" by how frequently known combinations of nucleons are found in nature. And by nature, the meaning here is only Earth.

For a reminder on weighted average see the _Calculating Weighted Averages_ section.

Notice that individual nucleon counts are whole numbers, but when averaged the results yield decimal numbers instead.

Additional names are known for the **Atomic Weight**:

- **Standard Atomic Weight,**

56

- Relative Atomic Mass, and
- Average Atomic Weight.

There are slight differences in the meanings between the Standard and Relative Atomic terms above, but within the scope of this book you can assume for now they are the same.

The Atomic Weight term does not possess the best wording, since it does not directly represent weight but rather atomic mass units. It is nevertheless strongly related to weight by providing the basis for calculating the actual weight mathematically.

Many periodic tables still contain the term Atomic Weight today, but it is being slowly phased out in favor of the more accurate "Relative Atomic Mass".

Elements have multiple isotopes and they find their way into samples of any chemical in typical ratios. Because it is not realistic to count the exact isotope composition of every sample, the only thing left to do is to estimate.

Scientists decide based on research how frequently different isotopes are to be expected. Each isotope gets a percent that determines how much it contributes to the global average. This calculation is given in the number represented by the periodic table's Atomic Weight (see *Figure 42 - Atomic Weight*).

Weight Uncertainties

Even though Atomic Weight estimations are based on thorough investigation, it is not difficult to imagine this method to carry some doubt.

Uncertainties about average weights are published by an organization called the *International Union of Pure and Applied Chemistry* (IUPAC). The first element to receive an **uncertainty in its atomic weight** was **Sulfur** in 1951[7].

The uncertainty can stem from strange localized combinations in samples from different geographic areas, labs, and manufacturers, and other factors.

The conclusion is that while common periodic table estimates are reasonable they cannot always be accurate for all samples.

Calculating Weighted Averages

The Atomic Weight is a **weighted** average of nucleons. To track calculations the following model is considered.

An atom represented by 3 equally frequent isotopes (weight variations) is shown next.

The isotopes differ by their nucleon counts. Nucleons are important as the only parts of the atom to embody matter and mass and thus their relationship to weight.

---

[7] Encyclopedia Britannica, Chemistry, and Physics,
https://www.britannica.com/science/atomic-weight

Electrons are not counted, they are basically energy and do not contribute weight to the matter.

Arithmetic Average

If the goal was to perform an arithmetic average for the next three isotopes, each then must be treated with equal importance. The average proceeds in the traditional manner and there are no additional considerations.

Three Variations of the Same Atom

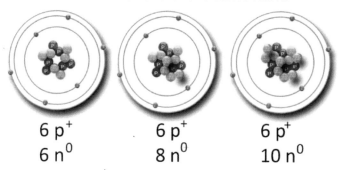

6 p$^+$ 6 p$^+$ 6 p$^+$
6 n^0 8 n^0 10 n^0

Figure 43 - The Average Weight of an Atom

The arithmetic average equals

$$\frac{\text{sum of all nucleons counts}}{\text{number of variations}}$$

In this example there are 3 variations and their nucleon counts (i.e. protons plus neutrons) from left to right are: 12, 14, and 16.

The average nucleon count in this set is thus

$$\frac{12+14+16}{3} = \frac{42}{3} = 14 \text{ \textbf{nucleons per atom}.}$$

Weighted Average

When the frequency of isotope occurrence is known, it must be taken into account. Rare isotopes should contribute less than those frequently found. For example, if 3 types of isotopes were found as follows:

n isotopes with	12 nucleons
m isotopes with	14 nucleons
k isotopes with	16 nucleons

then the next calculation gives the **weighted average** of the group.

$$\frac{n * 12 + m * 14 + k * 16}{n + m + k} \qquad \textit{(Equation 3)}$$

When replaced with actual numerical values

$$n=1, m=100, \text{ and } k=4,$$

one should be able to predict that m will have the greatest influence on the average as it is the most abundant. This also means that the weighted average should end up being relatively close to the count contributed by m, in this case 14 nucleons. The calculation is used to easily verify this idea.

$$\frac{1*12 + 100*14 + 4*16}{1+100+4} =$$

$$\frac{1*12 + 100*14 + 4*16}{1+100+4} = 14.057 \text{ nucleons.}$$

60

Giving percents instead of isotope amounts works the same way. Then n, m, and k represent percents that add up to 100% representing the entire group.

For example, if n=20%, m=50%, and k=30%, the sum of all percents is 20 % + 50 % + 30 % = 100 % and is seen as a full set.

The weighted average in this case becomes

$$\frac{20 * 12 + 50 * 14 + 30 * 16}{20 + 50 + 30} =$$

$$\frac{1,420}{100} = 14.20 \text{ nucleons.}$$

The same result is accomplished when the percents are listed in decimal form:

0.20 + 0.50 + 0.30 = **1.00**, indicating a full set as well. Simply replace the percents with their decimals,

$$\frac{0.20 * 12 + 0.50 * 14 + 0.30 * 16}{0.20 + 0.50 + 0.30} =$$

$$\frac{14.20}{1} = 14.20 \text{ nucleons.}$$

Atomic Mass Units

Review the earlier *Atomic Mass* section to correlate the total nucleon counts to the mass of the atom, or rather a specific isotope:

mass of isotope = the nucleon count = $p^+ + n^0$.

Every nucleon is equivalent to 1 unit of mass.

Transitioning from particle counts to mass units is immediate by extracting the **Symbol A** from the isotope notations. For example,

Carbon-12:   has 12 units of mass,

Helium-3:   has 3 units of mass,

Boron-11   has 11 units of mass.

Even though there are no differences between the nucleons living in separate atoms, to be safe a particular atom's isotope was chosen to contribute a nucleon as the chemical standard for mass. Namely this is the **Carbon-12** isotope.

Figure 44 - The Atomic Mass Unit (AMU) details the Carbon-12 transition from protons and neutron to the average nucleon, and furthermore to 12 average nucleons which span the entire mass of the isotope.

All nucleons have similar mass, though a neutron's mass is bit larger than the proton's. Averaging the mass of the two solves the problem.

Each nucleon represents 1 standard mass unit. With 12 of them present the conclusion is,

$$\textbf{1 unit of mass} = \frac{\text{the } Carbon-12 \text{ isotope's mass}}{12}$$

This standard unit of mass has been named the **AMU**. It stands for **"Atomic Mass Unit"** and in more recent years has been prefixed with the term "**unified**" and so is also known in short as **u** for "**unified atomic mass unit**", which is slowly taking over the **amu** notation. The two terms should be considered identical. Older books and most likely your professors are still using **amu**, newer documentation and many online resources are dropping the **amu** notation in favor of simply **u**. This book will use both interchangeably. **AMU, amu** and **u**, are also known as the **Dalton** unit designated in short by **Da**.

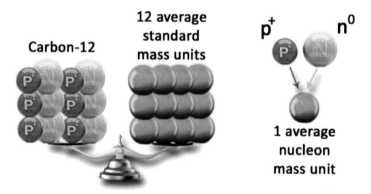

Figure 44 - The Atomic Mass Unit (AMU)

It should be also mentioned that other isotopes, like Oxygen-16, were once used instead as the standard, but Carbon-12 proved much easier to handle in the lab.

In conclusion,

> **1 amu = 1 nucleon of the Carbon-12 isotope.**

For now, the **amu** is the most precise way we know to describe the amount of matter the atom has.

Nevertheless, scientists can only deal with chemicals in units that make sense in their daily workflows and are measured easily. It is difficult to imagine the amu being the convenient unit for this, since it represents a particle count which is not realistic to work with. Instead of counting particles in samples, weighing them and designating amounts in grams, kg, etc., is a lot easier.

Before examining a correlation between the mass of one nucleon, and its weight in grams, let's clarify the difference between mass and weight.

Mass

The mass of an object is defined by the amount of matter present in it. Matter is the opposite of energy, it is tangible and gives the object density and dimensions which we can perceive and measure.

Mass is contributed by mass carrying particles such as protons and neutrons (see also potential mass alternates such as Dark Matter[8]). The mass of atoms is measured in atomic mass units, or AMU-s, and 1 mass unit is equivalent to a twelfth of the Carbon-12 mass.

In general, especially when relating to entities that are macroscopic, mass is exchangeable with weight and is measured in typical weight units such as g, mg, kg, etc.

---

[8] Dark Energy, Dark Matter/What Is Dark Matter?
https://science.nasa.gov/astrophysics/focus-areas/what-is-dark-energy

Weight

Weight was discovered in ancient times using scales which determined how much of a thing one had, and how it compared to other known weights.

Weight is different than mass. It is based on the pull an object's mass exerts downwards when hanging from a height. The drag is powered by **Gravity** and manifests as movement towards the planet's center of gravity that resides in its core.

The heavier an object is, the stronger the gravitational pull is too.

Gravity derives from matter and mass. It is dependent on several things:

(i)   the mass of the object being pulled,

(ii)   the mass of the planet that is creating the pull.

(iii) the distance at which the object resides in space, i.e. how far it is located from the center of gravity.

Figure 45 - Gravity Pulling Weight

For example, an astronaut is lighter on Mars than he/she is while spending time on Earth. A larger planet like Earth drives much greater Gravity than the smaller Mars on the same person possessing a steady mass.

Mass Weight Comparison

Mass	Weight
Is an absolute value.	Does not have a universally fixed value, depends on location.
Equals the total number of protons and neutrons.	It depends on the mass the object has.
	Varies with the object's location relative to the center of a Gravity such as Earth, Mars, etc.
Is measured in **amu**, or mg, g, kg, etc.	Is measured in mg, g, kg, etc.

Correlating the Mass and Weight of Nucleons

The correlation between

- the **mass** of one **nucleon**, or 1 AMU, and
- one nucleon's **weight** in grams,

was accomplished experimentally.

Figure 46 - Converting 1 amu to grams

To establish a conversion rate between the two, the **Carbon-12** isotope was chosen. It is the most stable and common isotope of **Carbon.** It is easy to handle in a lab so its overall **Atomic Mass** could be weighed accurately. Carbon-12 has the advantage of an even division of mass between its 6 protons and 6 neutrons. This is to ensure the average won't overrepresent one type of nucleon more than the other.

Carbon - 12

$6\,p^{+}$
$6\,n^{0}$

Figure 47 - Carbon-12

Once the weight of C-12 atom was measured, calculating the weight of a single nucleon was a matter of mathematical calculations, namely dividing the measurement by the 12 nucleons.

The following is a twelfth of the Carbon-12 weight measurement in grams,

1 amu = 1.6605 x 10^{-24} g. *(Equation 4)*

67

This provides a straight forward conversion between **Atomic Weight** and all weight units such as g, kg, etc.

Note: Read the _Math Review: Exponents_ section if you have difficulty with the 10^{-24} representation.

The steps below estimate the weight of a sample for a specific element.

1. The **Atomic Weight** of the element is pulled from the periodic table (see for example _Figure 48 - Atomic Weight of Neon_). It gives the average number of nucleons for the specified atom.
2. The above value is multiplied by the approximate number of atoms in the sample. The result is the average number of nucleons (or amu-s) in the sample.
3. AMU-s are converted into grams by multiplying the 1.6605×10^{-24} g conversion factor. The result is the estimated weight of the sample.

The method is useful for samples in large, macroscopic amounts. The **macroscopic** scale is the scale visible to the human eye, which is how we most frequently perceive and handle chemical compounds.

When the number of atoms **N**, in a chemical substance is known, the following method is used to calculate the **mass** of the substance in amu-s,

$$\text{mass} = N \times (\text{Atomic Weight}) \text{ amu}.$$

Then based on the rules listed above, the following product calculates the **weight**,

$$\text{weight} = N \times (\text{Atomic Weight}) \times 1.6605 \times 10^{-24} \text{ g.}$$

Example

Imagine an experiment in which it is known that a certain volume of Neon gas contains exactly 100,000 atoms. The **mass** of that volume can be calculated by:

$$N \times (\text{Atomic Weight}) =$$

100,000 x Atomic Weight of Neon in amu

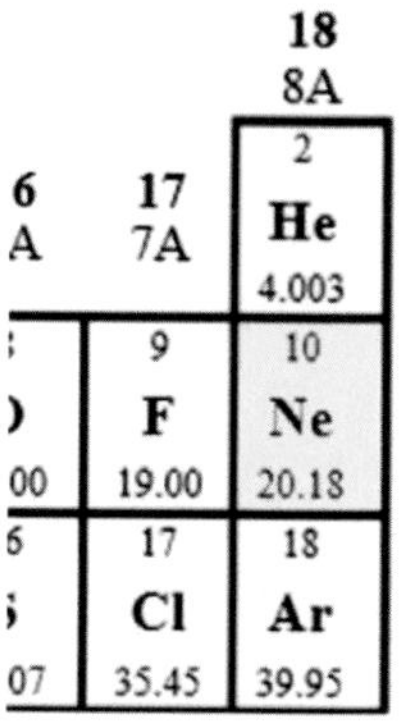

6 A	17 7A	18 8A
		2 He 4.003
;	9 F 19.00	10 Ne 20.18
00		
6 ; 07	17 Cl 35.45	18 Ar 39.95

Figure 48 - Atomic Weight of Neon

The atomic weight of Neon is read from the periodic table as **20.18 amu**. After plugging it into the calculation we get a mass of 100,000 x **20.18 amu = 2,018,000 amu** for this volume of gas.

Next, the result is converted from amu to grams to produce the **weight**.

2,018,000 amu x (1.6605 x 10^{-24} g per amu), and **g per amu** is expressed mathematically as $\frac{g}{amu}$.

2,018,000 amu x 1.6605 x 10^{-24} $\frac{g}{amu}$

Simplifying Units

Simplifying units is easiest when grouping numbers and units apart and treating them as separate expressions. Units of each term are shifted to the right side. Terms are recognized for being connected by additions and subtractions. When shifting units so, their original operators: multiplications and divisions, are carried too. From there, units simplify identically to mathematical variables. Therefore, the above expression becomes:

2,018,000 **amu** x 1.6605 x 10^{-24} $\frac{g}{amu}$ =

2,018,000 x 1.6605 x 10^{-24} ~~amu~~ x $\frac{g}{\text{~~amu~~}}$ =

2,018,000 x 1.6605 x 10^{-24} g = 3,350,889 x 10^{-24} g

Next, use up part of the exponent to adjust for rounding:

3,350,889 x $10^{-24-4+4}$ g = 3,350,889 x 10^{-4} x 10^{-24+4} g = 335.0 x 10^{-20} g.

In conclusion, a volume of **100,000 atoms** of Neon gas weighs approximately **335.0 x 10^{-20} g**.

Particle Mass Differences

When working with the amu unit, it must be remembered that it is based on the nucleon of a specific atom and is not a generic nucleon.

1 amu = 1 nucleon of the Carbon-12 isotope.

When the one proton of a Hydrogen mass is measured using mass spectrometry, the mass of the proton originating from the Hydrogen atom was found to be

mass of a Hydrogen's proton = 1.00727647 amu.

In other words, the mass of this specific proton exceeds 1 amu.

Hydrogen has only 1 proton and no neutrons. But there is another variation of the Hydrogen atom that does contain one neutron in addition to its proton. The atom is called Deuterium, also known as the ^{2}H isotope.

Two scientists in 1934, Chadwick and Goldhaber, have performed the first precise measurement of the neutron mass by looking at the photo-disassociation of the deuteron, i.e. the nucleus of Deuterium, and discovered that

mass of a ^{2}H neutron = 1.00866490 amu.

This creates a challenge since the average of a p^+ and n^0 should be 1 amu, but these measurements prove that both the proton and neutron mass is more than this average. It means one or both masses of p^+ and n^0 have strayed from their expected values in some way and their average would thus be larger than 1 amu.

More so, even if we were to accept these two new alternate values as the p^+ and n^0 masses, the math continues to not add up with experimental values. When measuring the mass of Deuterium as a combination of the two particles bound together as one nucleus, a.k.a. the deuteron, the expected mass would be naturally the sum of the two:

$$\text{expected mass}(^2H) = \text{mass of measured } p^+ + \text{mass of measured } n^0$$

$$= 1.00727647 \text{ amu} + 1.00866490 \text{ amu}$$

$$= 2.01594137 \text{ amu}$$

But instead of this value, the result was

measured mass(^{2}H) = 2.01410178 amu.

It puts the difference at a **loss of 0.00183959 amu** from where the expected weight should be.

Therefore, the mass of the two particles proton and neutron together weighs less than they do when they are apart. The mass lost is due to an effect called the Mass Defect.

Mass Defect

Mass Defect is a term used to describe the loss of mass resulting from stabilizing nucleons into a cohesive nucleus. The participating particles contribute bits of mass that get converted into energy at a rate based on Einstein's famous formula:

$$E = mc^2. \quad \textit{(Equation 5)}$$

The amount of energy lost to stabilization is also known as **Binding Energy**. If the nucleus was to be disassembled back into its subatomic particles, an identical amount of Binding Energy is released from the broken nucleic bind.

The energy returns to the individual nucleons that contributed it in the first place and converts back into mass. Experiments show that indeed the restored mass weighs more apart than it does when bound together.

Conclusion

The **amu** was defined explicitly in relation to the Carbon-12 mass and its own Binding Energy and Mass Defect.

The Mass Defect is different for every atom and has its individual stability requirements based on the number of nucleons present and other considerations.

Therefore, all atoms have inaccurate isotope masses listed in amu, except for Carbon-12. Values for mass are only averages for calculations but are proven to be good enough for predicting quantities of both reactants and products of chemical reactions.

Do We Know All Atoms That Can Exist?

Some 30 years ago, scientists listed fewer elements in the periodic table than they do today. Since then, many additional elements have been discovered.

The new elements were added mostly to the last two rows of the periodic table. They have high proton and neutron counts, making their nuclei large and heavy.

The latest atom to be added to the periodic table in 2016 was element numbered 118, once named **Ununoctium** but known today as **Oganesson**. It is the heaviest element known, and in spite of expectations for it to be a stable noble gas, it turned out to be a radioactive and extremely unstable solid. While the arrangement of the periodic table has predicted properties of new atoms accurately most of the time, there certainly are a few drastic exceptions such as Oganesson.

We do not know all the possible elements that may exist. New elements are being continuously discovered, or rather synthesized artificially for the last few decades.

Man-made elements are called **Synthetic Elements**. Approximately 24 of the synthetic elements can only exist in a lab for a split second before they naturally decay quickly by fracturing into smaller pieces. The so

75

shattered nuclei result in smaller groups of protons and neutrons that form their own lighter atoms as shown in the figure below. This process is called Nuclear Fission.

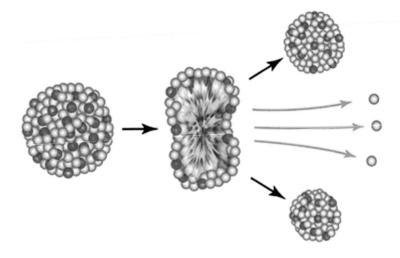

Figure 49 - Atomic Decay, Nuclear Fission

Chemistry came a long way in the last century. It is difficult to believe that it was only in 1905 that scientists first confirmed the existence of **atoms and molecules**. This was accomplished by French physicist and Noble laureate Jean Baptiste Perrin who based his experiments amongst other things on Einstein's work on Brownian Motion.

Counting Atoms

It is interesting to know how many atoms are present in any specific thing. It's easy to guess this number should be very large since atoms make up everything but they are invisible to the naked eye.

Let's take for example an average human cell. It is estimated to have approximately 100,000,000,000,000 atoms.

Colors were used to emphasize the magnitude of each portion of this number:

100, 000, 000, 000,000 atoms,

are equivalent to 100 trillion atoms.

100,	000,	000,	000,000
			million
		billion	
	trillion		

Such counts are very large involving the listing of numerous zeroes. It is not a convenient notation. One way to a more compact representation is to use exponents.

100 trillion in exponential form is equivalent to **10^{14}**.

Due to the use of extremely large numbers when counting atoms and molecules in chemistry, there is a strong need to engage notations that are more compact. This requires a good understanding of the **Exponent** as a mathematical concept (see the _Math Review: Exponents_ chapter at the end of this book).

A famous way of compacting numbers in science is called the **"Scientific Notation"**, which also requires the use of exponents.

Applying exponents is great but looking for additional methods of carving away some of these tremendous numerical magnitudes would be useful. The chapter below explores these possibilities.

Famous Amounts and Their Names

People have been frustrated in the past by the use of various quantities and opted to rename them. A few examples are shown.

1 pair	....	_2 pieces_
1 dozen	....	_12 pieces_
1 baker's dozen	....	_13 pieces_
1 myriad	....	_10,000 pieces_
1 googol	....	_10^{100} pieces_

Chemists deal with such and larger numbers in a similar way by renaming certain quantities.

Before examining what chemists have come up with, let's explore the mechanism of such a replacement model on an already known amount: the dozen.

Experiment with Dozens

Let's say a batch of **24 cookies** is available. The number of cookies is twice that of 12, i.e. the amount for one dozen. Instead of saying there are 24 cookies, one could say there are 2 dozen.

Since the trend is 12 cookies per 1 dozen, mathematically speaking the notation becomes:

$$\frac{12 \text{ cookies}}{1 \text{ dozen}} = 12 \frac{\text{cookies}}{\text{dozen}}$$

For a larger number of cookies, let's say **396**, the calculation is as follows.

$$\frac{396 \text{ cookies}}{12 \frac{\text{cookies}}{\text{dozen}}}$$

Units bundle together and follow their own math and simplify in the traditional way typical to numbers and variables. Numbers then appear on the left and units on the right.

For example, 100 cookies + 50 cookies = 150 cookies.

Before simplifying the units of the above example, fractions must be compacted.

$$\frac{396}{12} \text{ cookies} * \frac{1}{\frac{\text{cookies}}{\text{dozen}}}$$

We know from Math that fractions obey the rule:

$$\frac{1}{\frac{a}{b}} = 1 \div \frac{a}{b} = 1 * \frac{b}{a} = \frac{b}{a}$$

That is, division by a fraction is the same as the multiplication by that fraction's reciprocal.

$$\frac{396}{12} \text{ cookies} * 1 \frac{\text{dozen}}{\text{cookies}} =$$

$$33 * 1 \text{ cookies} * \frac{\text{dozen}}{\text{cookies}} =$$

After simplifying the units, the result is:

$$33 \quad \cancel{\text{cookies}} * \frac{\text{dozen}}{\cancel{\text{cookies}}} =$$

33 dozen.

Additionally, the 33 dozen can be converted back into cookies by multiplying according to the unit definition:

$$33 \text{ dozen} * 12 \frac{\text{cookies}}{\text{dozen}} = 33 * 12 \cancel{\text{dozen}} * \frac{\text{cookies}}{\cancel{\text{dozen}}} =$$

396 cookies.

To summarize what has been accomplished, our unit, the dozen, literally translates larger numbers into a renamed smaller dozen counterpart. That is, the expression 396 is replaced with the lesser 33 alternative. The larger the number of cookies to be translated, the more useful the conversion to dozen becomes.

Consider the enormous number of atoms and molecules in chemicals. Seems like the right place to use this technique. All that remains is to invent a new name and choose some clever amount that proves relevant for the chemical circumstance.

One such large number was introduced by the scientist **Jean Baptiste Perrin**, he called his number **Avogadro's Number**[9] to honor scientist Amedeo Avogadro who proposed a similar idea and founded the basics of the atomic-molecular theory.

The Mole

Initially Jean Baptiste Perrin (circa 1908) was interested in finding the number of atoms present in 1 gram of Oxygen. He called this potential count of atoms Avogadro's Number, which we know today as the mole. Later Oxygen was swapped out in favor of Carbon-12 and the calculations were adjusted accordingly.

The origin of the word mole is in Latin, meaning *"the amount of"*, and no, it's not that mole...

Mole

[9] https://www.scientificamerican.com/article/how-was-avogadros-number/

The Mole Definition

1 mole is defined as **6.022 x 10²³** units of anything, but typically, the units counted are: atoms, molecules, subatomic particles, etc.

This constant is known as **Avogadro's Number**, and is often denoted in formulas as N_A.

1 mol = 6.022 x 10²³ units = N_A	*(Equation 6)*

The word **mole** abbreviates as **mol**.

An alternate definition of the mole is **the number of atoms** that form a precise quantity of **12 g of the isotope Carbon-12.** It is notable that this amount is equivalent to a weight of 1 gram per nucleon of Carbon.

What to Count in Moles...

Amounts much smaller than 10^{23} are not necessarily counted in moles.

For example, the number of stars in our galaxy is a few hundred billion, a factor of $100 \times 10^9 = 10^{11}$. Even a **quadrillion** which equals 10^{15} is not large enough to have a meaningful expression in moles, as shown by the next calculation.

the number of moles in 1 quadrillion units =

$$\frac{10^{15}}{6.022 \text{ x } 10^{23}} = \frac{1}{6.022} * \frac{10^{15}}{10^{23}} = \frac{1}{6.022} * 10^{15-23} =$$

$$0.1661 \text{ x } 10^{-8} = 0.000000001661 \text{ moles.}$$

A previously mentioned example contemplated the 100 trillion atoms in the average human cell and whether its magnitude works well with the **mole** concept.

It is enough to look at the number's exponent to decide. The exponential form of 100 trillion is,

$$100 \text{ trillion} = 100{,}000{,}000{,}000{,}000 = 10^{14}.$$

The exponent, or magnitude of 100 trillion is 14, which is much smaller than 23, which is the exponent in Avogadro's number. The resulting count of moles would be similar to *0.00000...0n*, with approximately 23-14 = 9 zeroes after the decimal point.

By tracking the calculation that leads to the exact number of moles in 100 trillion (see *Math Review: The Rule of Three* one can better understand the ways in which the mole is to be used.

1 mole(s)   .... 6.022×10^{23} atoms

x moles   .... $100 \text{ trillion} = 10^{14}$ atoms

Before we begin the calculation, remember that units behave in fractions as if they were numbers or variables and thus should be simplified when possible. The placement of units in their expressions is implied to be like a multiplication. This is a convention that allows units to be moved commutatively so they can slide to the end of the expression.

For example,

$1 \text{ mol} * 10^{14} =$

$1 * 10^{14} \text{ mol}.$

Extracting x based on these proportions is easy for example using the *Rule of Three* method.

$$x \text{ moles} = \frac{1 \text{ mole(s)} * 10^{14} \text{ atoms}}{6.022 * 10^{23} \text{ atoms}}$$

Simplify the *atoms* unit.

$$x \text{ moles} = \frac{1 \text{ mole(s)} * 10^{14} \cancel{\text{atoms}}}{6.022 * 10^{23} \cancel{\text{atoms}}}$$

Next, slide the moles unit to the end of the expression.

$$x \text{ moles} = \frac{1}{6.022} * \frac{10^{14}}{10^{23}} \text{ moles}.$$

Verifying that the units of the final result are correct should always be part of one's error detection routine. For example, if the goal was to calculate a number of moles, but the final units cannot be simplified further and still relate to an inexplicable term, it is safe to assume an error made its way into the computation.

After applying the "A Power Divided by Another Power" exponent law, the result for x becomes:

$$x = 0.1661 * 10^{14-23} \; mol$$

$$x = 0.1661 * 10^{-9} \; mol$$

x can now be written also as

84

$x = 0.0000000001661 \ mol.$

Conclusion

Amazingly, 100 trillion atoms are nowhere near being close to even a single mole.

The mole was designed to tackle tremendously larger counts and significantly compact the notation. When applied incorrectly, it has the opposite effect and makes numbers look clunky. Thus, atoms in biological cells should not be counted in moles. Luckily, in chemistry you will probably not be asked to decide whether to count in moles or not, subatomic counts are nearly always counted in moles regardless of value, and most problems will state if different units are required.

Molar Mass

The Weight of 1 Mole

Whatever the weight of one mole of a specific element, we know it will contain precisely **6.0221 x 10²³** atoms. The question becomes how a mole of one atom is different than one mole of another.

Assume two atoms were selected to form mole groups and are shown in the figure below. On the left, atoms have 4 nucleons, and on the right 16 nucleons. Atomic counts are the same, each of the two groups contains exactly 1 mole of atoms, i.e. N_A = 6.022 x 10²³ units.

The 4:16 ratio of nucleons indicates each atom on the left is 4 times lighter than any one atom on the right.

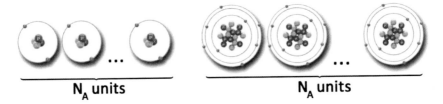

N_A units N_A units

Figure 50 - Comparing Moles

The total mass of each group is

$$M_1 = N_A * 4 \text{ amu} \qquad \& \qquad M_2 = N_A * 16 \text{ amu}$$

Notice,

$M_2 = N_A * 16 = 4 * (N_A * 4) = 4 * M_1,$

$M_2 = 4 * M_1,$

thus, if the individual atoms are 4 times heavier, then

the total weight of the group must also be 4 times heavier.

Based on *(Equation 4)* we can convert the amu-s to grams:

$W_1 = M_1 \times 1.6605 \times 10^{-24}\,g$

$W_2 = 4 \times M_1 \times 1.6605 \times 10^{-24}\,g$

From these calculations it becomes evident that the weight of 1 mole is not fixed, and it differs depending on which atom was selected.

Molar Mass Definition

The weight measured for 1 mole of a specific atom is called the atom's **Molar Mass**. For this, a special unit called **grams per mole** ($\frac{g}{mol}$) was chosen.

The Molar Mass definition seeks to find - just like its unit implies - the quantity in grams of all atoms counted as part of 1 mole. The figure below captures this idea visually.

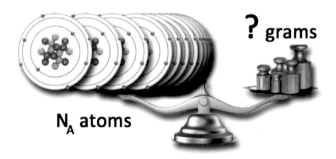

? grams

N_A atoms

Figure 51 - Grams per Mole

An atom's Molar Mass, i.e. the mass of 1 mole will depend on the type of atom selected. Heavier atoms (on the right) have a heavier mole.

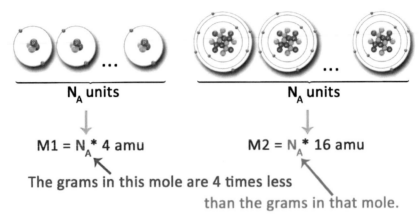

N_A units N_A units

M1 = N_A * 4 amu   M2 = N_A * 16 amu

The grams in this mole are 4 times less

than the grams in that mole.

Figure 52 - Comparing Molar Masses

Molar Mass is often used in chemical expressions and appears in short as the capital letter M. For example, the molar mass of Magnesium and the molar mass of Oxygen can be designated as **M(Mg)** and **M(O)** respectively.

Molar Mass could easily be expressed in *amu per mole* instead of *grams per mole*, since this is after all a mass equivalent translated via a total of nucleons in the atom. However, it is hard to imagine a workflow in which one would know how much of a substance to provide when told to prepare some number of nucleons.

Instead, it makes more sense to prepare a quantity in grams which is then easy to supply by weighing the substance on a scale. The next section examines how this idea is accomplished.

Weight Calculations

Previous chapters have demonstrated that calculations from **amu** to **grams** are straight forward. The conversion rate is given in section _Correlating the Mass and Weight of_ Nucleons. The factor is based on the Carbon-12 isotope. These were the realizations:

1 average nucleon = 1 unit of mass = 1 amu,

and the weight of a single C-12 nucleon is

1 amu = 1.6605 x 10^{-24} g.

It is worth noticing the mathematical relationship between the reciprocal of 1 amu weight in grams and Avogadro's number is one of equality:

$$\frac{1}{1.6605 \text{ x } 10^{-24}} = 6.0221 * 10^{23} = N_A$$

The mass of 1 nucleon can be remembered as the reciprocal of Avogadro's number:

$$1 \text{ amu} = \frac{1}{N_A} \text{ g.} \quad \textit{(Equation 7)}$$

Negligible Mass Errors

Describing the weight of an atom in amu is just as correct as describing it in grams. Converting back and forth between the two is generally considered accurate, though mathematically conversion errors are to be expected. This stems from the nucleon mass being averaged from protons and neutrons, neutrons having a bit more mass and most isotopes possessing a lot more neutrons than protons. Larger atoms thus have extra neutron-mass unaccounted for by the amu unit. This doesn't happen when there is an equal proton and neutron count.

Mole Calculations

The chosen standard of mass for mole weight calculations is the nucleon of the Carbon-12 isotope. The nucleon is the average of exactly one proton and one neutron mass originating from an even 6:6 protons to neutrons mass ratio.

To relate atomic counts to measurable quantities in grams, one question of interest is: **How many C-12 atoms are needed to measure a quantity of 1 gram?**

Figure 53 - Number of Atoms in 1 gram

90

A quick review of what we know so far should help:

- 1 mole = N_A = 6.0221 x 10^{23} instances of anything,
- the mass of 1 nucleon represents 1 amu, and it weighs 1.6605 x 10^{-24} g = $\frac{1}{N_A}$ g,
- the number of nucleons in one C-12 atom is 12;
- counting 1 nucleon is the same as counting $\frac{1}{12}$ -th of a Carbon-12 isotope.

Read *Math Review: The Rule of Three* at the end of this book if the proportions below are confusing.

The goal is to solve for x, where

x = the number of Carbon-12 atoms in 1 gram.

1 gram of C-12	....	x atoms of C-12
1.6605 x 10^{-24} g	....	$\frac{1}{12}$ atoms of C-12

$$x = \frac{1\,g * \frac{1}{12}\,\text{atoms}}{1.6605 * 10^{-24}\,g}$$

Simplify the units where applicable, in this case the grams. Notice all values in the expression are multiplied, there are no additions or subtractions, so simplifying is straight forward:

$$x = \frac{1\,\cancel{g} * \frac{1}{12}\,\text{atoms}}{1.6605 * 10^{-24}\,\cancel{g}}$$

We know from Mathematics that

$$\frac{\frac{a}{b}}{c} = \frac{a}{b} * \frac{1}{c},$$

Thus, $\frac{1}{12}$ can be place in the front:

$$x = \frac{1}{12} * \frac{1}{1.6605 * 10^{-24}} \text{ atoms}$$

$$x = \frac{1}{12} * \frac{1}{\frac{1}{N_A}} \text{ atoms}$$

$$x = \frac{1}{12} * N_A \text{ atoms}$$

The conclusion is that x, the number of atoms **in one gram** of Carbon-12 is one twelfth of Avogadro's number. In which case, 12 grams would have to be 12 x atoms, which in turn yield the full value for Avogadro's number.

$$12 * x = N_A \text{ atoms} \quad \text{(Equation 8)}$$

This calculation reveals why the mole definition was adapted to the number of atoms in precisely 12 g of Carbon-12.

Did it Have to Be Carbon-12?

In the early 1900s it was Oxygen, not Carbon that was the standard for mole calculations. So, definitely the standard did not have to be Carbon-12.

If one was to reestablish Oxygen to its former glory for measuring both the **amu** unit and the mole, the same N_A count would still be achieved. However, atomic counts would not be a match to 12 g of atoms, but rather to 16 g, because Oxygen has 16 nucleons. Yet the question remains the same: how many atoms are there in 1g? The details of that calculation transform so:

1 gram of Oxygen-16 ...... x atoms of O-16

$$1.6605 \times 10^{-24} \text{ g} = \frac{1}{N_A} \text{ g} \quad \cdots\cdots \quad \frac{1}{16} \text{ atoms of O-16} =$$

$$= 1 \text{ nucleon} = 1 \text{ amu}$$

$$x = \frac{1 * \frac{1}{16}}{\frac{1}{N_A}} = \frac{1}{16} \div \frac{1}{N_A} = \frac{1}{16} * \frac{N_A}{1}$$

$x = \frac{1}{16} * N_A$, multiply both sides of the equation by 16.

$16 * x = N_A$ atoms,

where x is the number of atoms in 1 gram of Oxygen-16.

The resulting number of atoms in 1 mole is the same as before. However, in the case of Oxygen the mole definition would have to be modified to

"the mole equals the number of atoms present in 16 grams of Oxygen, rather than the 12 grams for Carbon",

to reflect the new standard.

The increase in grams makes sense considering that the C-12 isotope is lighter than O-16 due to fewer of its nucleons. Then comparing the two sets of N_A atoms (see *Figure 52 - Comparing Molar Masses*), makes it clear the Carbon-12 group will weigh 12 g, and the heavier group of Oxygen-16 in an identical count of atoms weighs 16 g.

This confirms the earlier idea that a mole of different atoms will weigh differently, but the mole always counts the same number of atoms.

Turns out the standard did not have to be Carbon-12. Using another atom would designate a different weight for 1 mole. There are several atoms that could be chosen for the calculation of the mole, though not all are equally convenient to handle.

Remembering that Carbon-12 is the current mass/weight standard is an important general knowledge trivia that allows the mole calculation to be tracked with ease.

Trends
While exploring different possible standards, a trend has emerged between

- x_i = **the number of atoms in 1 gram of isotope,**
- the overall number of nucleons present, and
- N_A = Avogadro's Number.

Isotope Notation	Isotope Hyphenated-Notation	Atomic Mass or nucleons	Trend
^{12}C	Carbon-12	12	$12 * x_C = N_A$
^{16}O	Oxygen-16	16	$16 * x_O = N_A$
^AX	generic isotope X-A	A	$A * x_X = N_A$

If ^AX is the isotope of a generic atom X containing A nucleons, then:

A = the **Atomic Mass** of isotope ^AX, and

x_X = the number of X atoms in 1 gram of isotope.

In reality, 1 gram of substance will never contain only one type of isotope of the parent atom unless specifically engineered so on purpose. Instead, a mixture of isotopes is always present. The Atomic Mass is replaced by the average of all isotope Atomic Masses. The average is called the **Atomic Weight**. The original formula of the trend:

$$A * x_X = N_A,$$

becomes

$$\text{Atomic Weight} * x_X = N_A.$$

To explore this further, assume a substance was given after weighing as a mass called **m**, measured in grams.

95

Assume the mass has a total number of nucleons **p** distributed amongst all its atoms. One nucleon weighs

1 amu = 1.6605 x 10^{-24} g, and thus

$$\frac{m \ grams}{1.6605 \ x \ 10^{-24} \ grams} = \textbf{p} \text{ nucleons.}$$

$$\textbf{p} \text{ nucleons} = \frac{m \ grams}{1.6605 \ x \ 10^{-24} \ grams}$$

The mass of one nucleon and Avogadro's Number are reciprocals:

$$1.6605 \ x \ 10^{-24} = \frac{1}{N_A} = m_{nucleon}, \text{ see } \textit{(Equation 7)}$$

$$\textbf{p} \text{ nucleons} = \frac{m}{\frac{1}{N_A}}$$

$$\textbf{p} \text{ nucleons} = m \div \frac{1}{N_A} = m * \frac{N_A}{1}$$

p nucleons = $m * N_A$

The number of nucleons **p** belongs to identical isotopes. Each isotope has an Atomic Mass **A**, i.e. there are **A** nucleons in each atom. Assume the number of atoms is called **k**, then the relationship between **k**, **p** and **A** is:

$$\textbf{k} \text{ atoms} = \frac{p}{A}$$

96

After replacing **p** with $m * N_A$ we get:

$$k \text{ atoms} = \frac{m * N_A}{A}$$

Once more the assumption must be that a mixture of isotopes is always received with every batch of atoms rather than identical isotopes. Then a more general approximation of real life quantities must replace **A** with the **Atomic Weight** average.

$$k \text{ atoms} = \frac{m * N_A}{\text{Atomic Weight}}$$

It is better to count atoms in moles rather than atoms. Converting **k** atoms into a number of moles **n** is easy. Remember that N_A is the number of atoms in one mole, see *(Equation 6)*.

$$n \text{ moles} = \frac{k}{N_A},$$

and after plugging in the calculation for **k**, the conclusion is:

$$n \text{ moles} = \frac{\frac{m * N_A}{\text{Atomic Weight}}}{N_A} = \frac{m * N_A}{\text{Atomic Weight}} \div \frac{N_A}{1}$$

$$n \text{ moles} = \frac{m * N_A}{\text{Atomic Weight}} * \frac{1}{N_A}$$

$$n \text{ moles} = \frac{m * N_A}{\text{Atomic Weight}} * \frac{1}{N_A}$$

$$n \text{ moles} = \frac{m}{\text{Atomic Weight}},$$ and solving for **m** yields:

m = n * Atomic Weight.

The **Atomic Weight**, an average of all expected isotopes' Atomic Masses is read directly from the periodic table as follows.

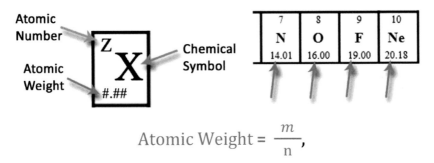

$$\text{Atomic Weight} = \frac{m}{n},$$

To differentiate between

- a proper Atomic Weight, measured in amu, and
- an atomic weight given by the above calculation but measured in **grams per mole** $(\frac{g}{mol})$,

the term Molar Mass, or in short M, was introduced and replaces the traditional Atomic Weight:

98

M = Atomic Weight in $\frac{g}{mol}$ units, instead of amu.

The formula is rewritten as

$$M = \frac{m}{n},$$

and represents a fundamental notion in a branch of chemistry called **Stoichiometry**. It is indispensable in all calculations relating to quantities in chemical reactions.

The difference between **M** and the **Atomic Weight** is only a matter of units and no conversion is necessary.

To finalize the idea, if

Atoms Weight = **w** amu,

then the atom's corresponding **Molar Mass** is

$$M = w \ \frac{g}{mol}. \quad \text{(Equation 9)}$$

Molar Mass and Chemical Formulas

Molar Mass is applicable to Chemical Formulas the same way as it is to individual atoms.

A **Chemical Formula** describes a substance more complex than a single atom. It shows a combination of individual or groups of atoms.

There are several ways to write these formulas, below are a few examples:

Glucose $C_6H_{12}O_6$

99

Manganese(IV) Sulfate $\quad\quad$ $Mn(SO_4)_2$

Acetic Acid (vinegar) $\quad\quad$ CH_3COOH

The things to notice here are:

- The atoms and their counts are listed in the chemical formula as chemical symbols and numerical subscripts.

- Entire groups can repeat together and contribute their number of atoms accordingly. For example, $(SO_4)_2$ is the same as $(SO_4) + (SO_4)$, in other words 2 S atoms, and 8 O atoms are contained.

- Each atom's Molar Mass can be looked up from the periodic table.

- The overall Molar Mass of a chemical formula is computed by adding up all molar masses of its individual atoms.

The equation

$$M = \frac{m}{n}, \quad \textit{(Equation 10)}$$

can also be expressed as

$$m = M * n. \quad \textit{(Equation 11)}$$

These relations are used with chemical formulas not just individual atoms. They correlate mass, moles, and overall Molar Mass for a given substance.

To designate the Molar Mass of a specific element the **M(element)** notation is used. Similarly, the Molar Mass of a chemical formula appears as **M(formula)**, as shown in the next instance.

Example

This example demonstrates the calculation of Molar Mass for vinegar: CH_3COOH.

The chemical formula for vinegar can be written more compactly as $C_2H_4O_2$ to summarize overall atomic counts. The formula has 2 Carbon atoms, 4 Hydrogens, and 2 Oxygens.

$$M(C_2H_4O_2) \quad = M(C_2) + M(H_4) + M(O_2)$$
$$= 2M(C) + 4M(H) + 2M(O)$$

Molar Masses of individual atoms are read from the periodic table and plugged into the calculation.

1 1A								18 8A
1 **H** 1.008	2 2A		13 3A	14 4A	15 5A	16 6A	17 7A	2 **He** 4.003
3 **Li** 6.941	4 **Be** 9.012	•••	5 **B** 10.81	6 **C** 12.01	7 **N** 14.01	8 **O** 16.00	9 **F** 19.00	10 **Ne** 20.18
11 **Na**	12 **Mg**		13 **Al**	14 **Si**	15 **P**	16 **S**	17 **Cl**	18 **Ar**

$M(C) = 12.01$, $M(H) = 1.008$, $M(O) = 16.00$

Thus,

$$M(C_2H_4O_2) = 2M(C) + 4M(H) + 2M(O)$$
$$= 2*12.01 + 4*1.008 + 2*16.00$$
$$= 24.02 + 4.032 + 32.00$$
$$= 60.05$$

$M(C_2H_4O_2) = 60.05$.

Atomic Theory Quiz

(see also: *Atomic Theory Quiz,* Answer Key)

1. Are atoms and elements the same thing?
   a) Yes
   b) No

2. Most things on Earth are made of
   a) only one element
   b) many atoms combined together
   c) neither

3. What are atoms made of?
   a) Lego shaped boxes
   b) subatomic particles
   c) smaller atoms
   d) scientists don't know yet

4. Which of the following parts are located inside atoms? (select all that apply)
   a) nucleus
   b) p^+
   c) e^-
   d) neutrons

5. The nucleus is made of
   a) p^+
   b) e^-

103

c) e⁻, p⁺ and neutrons

d) neutrons

e) e⁻ and neutrons

f) p⁺ and neutrons

6. The nucleus is kept together by
    a) the Weak Force
    b) the Strong Force
    c) Electromagnetism
    d) Static Electricity

7. The "Solar System Model" for atoms is used to simplify what concept?
    a) the nucleus
    b) the p⁺ gluing to neutrons
    c) the e⁻ orbits

8. Which of the following particles have mass? (select all that apply)
    a) p⁺
    b) neutrons
    c) e⁻
    d) e⁻ orbits

9. Which of the following particles are mostly energy? (select all that apply)
    a) p⁺
    b) neutrons

c) e⁻

d) e⁻ orbits

e) neither

10. Another name for the Strong Force is

 a) Static Charge

 b) Electromagnetism

 c) Strong Interaction

 d) Electrostatic Repulsion

11. Which combinations of particles are known to repel? (select all that apply)

 a) p^+ & neutrons

 b) p^+ & p^+

 c) e⁻ & e⁻

 d) e⁻ & p^+

 e) e⁻ & neutrons

12. Which combinations of particles are known to attract? (select all that apply)

 a) p^+ & neutrons

 b) p^+ & p^+

 c) e⁻ & e⁻

 d) e⁻ & p^+

 e) e⁻ & neutrons

13. Particles repel due to

 a) the Strong Force

b) Electromagnetism
c) the Weak Force

14. Which particle is neutral?
    a) p^+
    b) neutron
    c) e^-
    d) e^- orbits
    e) neither, all particles are charged

15. Which are examples of variations of the same atom?
    (select all that apply)
    a) Atomic Forces
    b) Electrons
    c) Isotopes
    d) Ions
    e) Atomic Bonds
    f) Protons
    g) Water molecules

16. We can tell atoms apart by the number of
    a) p^+
    b) neutrons
    c) e^-
    d) orbits
    e) bonds

17. A mole is

106

a) an animal used by Chemists in experiments
b) the birthmark of a famous Chemist
c) a counter similar to: pair, dozen, baker's dozen
d) any chemical found underground

18. Use pen and paper to calculate how many moles are contained in $150.55 * 10^{25}$ atoms?
(Reminder: 1 mole is equivalent to 6.022×10^{23} pieces.)
a) 25 moles
b) 350 moles
c) 2,500 moles
d) 10,678 moles

19. Should photons be counted in moles?

a) Yes
b) No

20. What is the total count in 21 moles of photons?

a) 6.022×10^{23} photons
b) 126.5×10^{27} photons
c) 0.2868×10^{23} photons
d) 1.265×10^{25} photons

Atomic Theory Quiz, Answer Key

1. a

2. b

3. b

4. a, b, c & d

5. f

6. b

7. c

8. a & b

9. c

10. c

11. b & c

12. d

13. b

14. b

15. c & d

16. a

17. c

18. c

Solution:

Use the Rule of Three to align the proportions correctly.

	1 mole is to	6.022×10^{23} atoms
like	x moles are to	$150.55 * 10^{25}$ atoms.

Apply the Rule of Three.

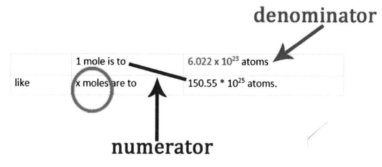

$$x \text{ moles} = \frac{1 \text{ mol} * 150.55 * 10^{25} \text{ atoms}}{6.022 * 10^{23} \text{ atoms}} =$$

$$\frac{150.55 \text{ mol} * 10^{25} \text{ atoms}}{6.022 * 10^{23} \text{ atoms}} =$$

$$\frac{150.55 \text{ mol} * 10^{25} \text{ atoms}}{6.022 * 10^{23} \text{ atoms}} =$$

$$\frac{150.55 \text{ mol}}{6.022} * \frac{10^{25}}{10^{23}} =$$

$$25.00 \text{ mol} * 10^{25-23} =$$

$$25.00 \text{ mol} * 10^{2} =$$

$$25.00 * 100 \text{ moles}$$

In conclusion:

$x = 2{,}500$ moles.

19. a

20. d

Solution:

Total photon count = 21 moles of photons =

$21 * (6.022 * 10^{23})$ photons =

$21 * 6.022 * 10^{23}$ photons =

$126.5 * 10^{23}$ photons =

$1 * 126.5 * 10^{23}$ photons,

and $1 = \dfrac{10^{-2}}{10^{-2}}$

$\dfrac{10^{-2}}{10^{-2}} * 126.5 * 10^{23}$ photons =

$10^{-2} * 126.5 * \dfrac{10^{23}}{10^{-2}}$ photons =

$1.265 * 10^{23-(-2)}$ photons =

$1.265 * 10^{25}$ photons

The answer is there are $1.265 * 10^{25}$ units in 21 moles of photons.

Exercises

Question 1

Give the name of the elements with the following atomic numbers: 6, 7, 8, 10, 36, 54.

Question 2

How many neutrons are there in the following isotopes: ^{18}F, ^{18}Ne, ^{37}Cl, ^{40}Ar?

Question 3

Express the atomic mass of one Germanium ^{70}Ge isotope as a fraction of the ^{12}C isotope's mass.

Question 4

The definition of the mole is: $N_A = 6.0221$ x 10^{23} units, and it is equal to the number of atoms contained in 12 grams of ^{12}C.

How many grams of ^{28}Si would be necessary if the standard of mass for the mole calculation was to be swapped from ^{12}C to ^{28}Si?

Question 5

Calculate the weight in grams of 10^{32} atoms of Neon gas.

Question 6

How many moles of Mg atoms are contained in 25 g of Magnesium?

Question 7

Calculate the molar mass of glucose: $C_6H_{12}O_6$.

Subscripts indicate how many of each atom are present in the given chemical formula, in this case 6 Carbons, 12 Hydrogens, and 6 Oxygens.

Question 8

Calculate the molar mass of Manganese(IV) Sulfate: $Mn(SO_4)_2$.

Here too, and this will always be the case, subscripts indicate how many of each atom are present in the chemical formula given, i.e. 1 Manganese and two groups of SO_4.

Make sure to notice that each group of SO_4 contains one Sulfur atom and 4 Oxygens.

Question 9

How many grams of SiO_2 are contained in one mole of this substance?

Question 10

How many grams of O_2 are present in 0.0125 moles of this molecule?

Question 11

Imagine X to be an atom. It occurs naturally in the following isotopes and their respective abundance.

112

isotope	abundance
X-28	92.21%
X-29	4.70%
X-30	3.09%

What is the Atomic Weight of X?

Question 12

What element is X in the previous question?

Question 13

List the number of protons and neutrons in the ^{30}Si isotope. Use the periodic table to help you answer this question.

Question 14

(a) How many moles are there in 100.00 g of sulfuric acid H_2SO_4?

(b) How many total moles of atoms are present in the above amount?

Question 15

(a) How many moles are there in 500.00 g of CO_2? Before you answer the question, what do you think the moles in this question refer to: atoms or molecules? Explain why.

(b) How many molecules are there in the above amount? Notice the question is not moles of molecules but simply molecules.

Question 16

There is a mystery atom X in the $X(NO_3)_2$ chemical formula. The compound containing X is given in a quantity of 15.02 g, and its equivalent number of moles is 0.0915 mol. What is element X?

Question 17

(a) What is the molar mass of fructose? The chemical formula for the fructose molecule is: $C_6H_{12}O_6$. Fructose is a type of sugar found in fruits, flowers, and various other plants.

(b) An equivalent notation for fructose is $(CH_2O)_6$.

Other similar molecules are formatted as $(CH_2O)_x$, where **x** is seen as the number of **repetitions** of the inner base group CH_2O. For fructose x is 6, other molecules have their own versions of x. How many repetitions x should be allowed for a $(CH_2O)_x$ molecule so that it has a molar mass of $60.06 \dfrac{g}{mol}$?

(c) Research the name of the molecule formed by the number of repetitions x you found in Part (b).

Solutions to Exercises

You will need a periodic table for the solutions listed below. Periodic tables are changing frequently in tandem with the latest research. For an updated version try the *Royal Society of Chemistry* interactive periodic table at: http://www.rsc.org/periodic-table. It provides element information such as the name, Atomic Number, and much more. Note that table cells have minimal information until the computer mouse is hovered above it. Experiment with various elements, hover the mouse to see the Atomic Weight, main isotopes, density and more. To appreciate how the periodic table has changed over time, navigate to: http://www.rsc.org/periodic-table/history and select for example the year 2000, to spot which elements were unknown at that time.

Solution 1

Atomic numbers are typically listed at the top of the periodic table cells. The following atomic numbers have been colored below: 6, 7, 8, 10, 36, and 54.

6	7	8	9	10
C	N	O	F	Ne
12.01	14.01	16.00	19.00	20.18
14	15	16	17	18
Si	P	S	Cl	Ar
28.09	30.97	32.07	35.45	39.95
32	33	34	35	36
Ge	As	Se	Br	Kr
72.59	74.92	78.96	79.90	83.80
50	51	52	53	54
Sn	Sb	Te	I	Xe
118.7	121.8	127.6	126.9	131.3

The information extracted from the periodic table is listed in the next table.

Atomic Number	Symbol	Element Name
6	C	Carbon
7	N	Nitrogen
8	O	Oxygen
10	Ne	Neon
36	Kr	Krypton
54	Xe	Xenon

The typical periodic table does not list the names of elements, only their chemical symbol. To find the names of these elements use the *Royal Society of Chemistry* digital periodic table at http://www.rsc.org/periodic-table).

Solution 2

The isotopes listed in this question are written in the compact isotope notation:

Z is dropped, because the p⁺ count is implied by the chemical symbol **X** which clearly identifies the relevant element.

A is the number of nucleons and equal to the Atomic Mass, i.e. *(Equation 2)*:

$$A = p^+ \text{ count} + n^0 \text{ count}$$

The proton count is the same as the Atomic Number and is listed for every element in the periodic table. The number of neutrons can be derived by rearranging the above equation:

$$n^0 \text{ count} = A - \text{Atomic Number}$$

Fill out a table as follows to identify the information requested. First copy the four isotopes in question into the "*Isotope*" column. Next, copy the Atomic Mass listed in the isotope notation's top left corner into the "*A*" column.

Check a periodic table (for example: http://www.rsc.org/periodic-table) for the Atomic Number of each element, listed in the upper area of cells and add them to the "*Atomic Number*" column.

Calculate the difference between the last two columns and log the result in the "*n^0 count*" column as shown.

Isotope	A	Atomic Number	n^0 count = A − Atomic Number
^{18}F	18	9	18 - 9 = 9
^{18}Ne	18	10	18 - 10 = 8
^{37}Cl	37	17	37 - 17 = 20
^{40}Ar	40	18	40 − 18 = 22

Thus, the neutron counts in the following isotopes: ^{18}F, ^{18}Ne, ^{37}Cl, ^{40}Ar are: 9, 8, 20, and 22 respectively.

Solution 3

In general, every isotope contributes 1 amu mass for each nucleon it has. That's 12 for ^{12}C, and 70 nucleons for ^{70}Ge, as indicated by their Atomic Mass superscript in the isotope notation.

1 ^{12}C isotope   .... 12 amu

1 ^{70}Ge isotope   .... 70 amu

The question then is how many ^{12}C isotopes x are equivalent to 1 ^{70}Ge isotope, or rather 70 amu? This can be seen as mass, or as nucleon counts, both will yield the same result.

1 ^{12}C isotope   .... 12 amu

x ^{12}C isotopes   .... 70 amu

Solving for x results in:

$$x = \frac{1*70 \; amu}{12 \; amu},$$

Simplify the amu units,

$$x = \frac{70}{12} \cong 5.83 \text{ isotopes (or atoms).}$$

Thus, one ^{70}Ge atom weighs approximately as much as 5.83 Carbon-12 atoms, or in other words, one ^{70}Ge atom has approximately 5.83 times more nucleons than the ^{12}C isotope.

Solution 4

The solution to this problem is nearly identical to the problem proposed in section _Did it Have to Be Carbon-12?_ In this case we are dealing with ^{28}Si rather than the ^{16}O discussed in that chapter.

First, ^{28}Si must be detailed:

- the Atomic Number of Silicon is Z = 14 protons (read from the periodic table),

- it has an Atomic Mass of 28 nucleons, known from the isotope name: ^{28}Si.

- because there are 28 nucleons in the ^{28}Si isotope, 1 nucleon of Silicon mass = the weight of 1 Si isotope ÷ 28 parts, i.e. one 28th part of the total atom's weight.

We know the mass of one nucleon from the _Correlating the Mass and Weight of Nucleons_ section to be

$$1 \text{ nucleon mass} = 1 \text{ amu} = 1.6605 \text{ x } 10^{-24} \text{ g} = \frac{1}{N_A} \text{ g,}$$

see _(Equation 4)_ and _(Equation 7)_.

Then the following equivalence can be noted:

1 gram of Silicon-28 .... x atoms of Si-16

1.6605×10^{-24} g $= \dfrac{1}{N_A}$ g .... $\dfrac{1}{28}$ atoms of Si-16

Extracting x from these proportions is based on the Rule of Three:

$$x = \dfrac{1 * \dfrac{1}{28}}{\dfrac{1}{N_A}},$$

$$x = \dfrac{1}{28} \div \dfrac{1}{N_A} = \dfrac{1}{28} * \dfrac{N_A}{1},$$

$$x = \dfrac{1}{28} * N_A,$$

multiply both sides of the equation by 28,

$28 * x = N_A$ atoms, where x in this case is the number of atoms in 1 gram of Silicon-28.

The question was how many grams of ^{28}Si would be necessary to fit exactly 1 mole so that the standard of mass could be replaced from C-12 to Si-28. The above calculation confirms that a N_A count of Silicon atoms would be equivalent to

$$28 * (1 \text{ gram's worth of atoms} = x).$$

In other words, a N_A count of Silicon atoms would be equivalent to exactly 28 grams of ^{28}Si.

The answer to the question is thus 28 grams.

120

Solution 5

There are several ways to solve this. If it was known how many moles in 10^{32} atoms, let's call this number **n**, then makes sense to use the formula **m = n*M**. Where **M** is the Molar Mass in $\frac{g}{moles}$ adapted from the periodic table.

First Solution

First the moles represented by 10^{32} atoms are based on N_A = Avogadro's Number:

$$n = \frac{10^{32} \text{ atoms}}{N_A \frac{\text{atoms}}{\text{mol}}}$$

$$n = \frac{10^{32} \cancel{atoms}}{N_A \frac{\cancel{atoms}}{mol}}$$

$$n = \frac{10^{32}}{6.022 \text{ x } 10^{23} \frac{1}{\text{mol}}}$$

$$n = \frac{10^{32-23}}{6.022} \frac{\text{mol}}{1}$$

$$n = \frac{10^{9}}{6.022} \text{ mol} = \frac{1}{6.022} * 10^{9} \text{ mol}$$

$$n = 0.1660 * 10^{9} \text{ mol}$$

17	2
7A	He
	4.003
9	10
F	Ne
19.00	20.18
17	18
Cl	Ar

The Atomic Weight of Neon in the periodic table is 20.18 **u** or **amu**. We use this number as the Molar Mass but first adjust the units:

$$M(Ne) = 20.18 \frac{g}{mol}$$

The mass can be extracted from the formula:

m = n * M, see *(Equation 11)*, after plugging in the values we found so far.

$$m = 0.1660 * 10^9 \; mol * 20.18 \frac{g}{mol}$$

$$m = 0.1660 * 10^9 \; \cancel{mol} * 20.18 \frac{g}{\cancel{mol}}$$

$$m = 0.1660 * 20.18 * 10^9 \; g.$$

m = 3.34988 * 10^9 g, and after rounding the weight of 10^{32} Ne atoms is:

m = 3.350 * 10^9 g.

Second Solution

The second way is to remember that the mass of one nucleon is 1.6605 x 10^{-24} g.

The average Neon isotope has an Atomic Weight of 20.18 amu. From here the weight of 1 atom is calculated by:

$$20.18 \text{ amu} * 1.6605 \times 10^{-24} \frac{g}{amu}.$$

There are 10^{32} atoms to account for. The mass of all these atoms together is:

$m = 10^{32} * (20.18 * 1.6605 * 10^{-24})$ g

$m = 33.51 * 10^{32-24}$ g $= 33.51 * 10^8$ g

Adjust the exponent so the results from the two solutions can be compared with ease:

$m = 33.51 * 10^{-1} * 10^1 * 10^8$ g

$m = 3.351 * 10^1 * 10^8$ g $= 3.351 * 10^{1+8}$ g

$m = 3.351 * 10^9$ g.

The two results are close enough to be considered equal.

Solution 6

First find Magnesium in the periodic table. It is in the second column and third row.

Li	Be	
6.941	9.012	
11	12	
Na	Mg	
22.99	24.31	
19	20	21
K	Ca	Sc
39 10	40 08	44 96

The Molar Mass of this element is the same as its Atomic Weight but using the $\frac{g}{mol}$ units.

$$M(Mg) = 24.31 \frac{g}{mol}$$

(*Equation 11*) states:

$$m = n * M,$$

where

- **m** is the mass in grams,
- **n** is the number of moles, and
- **M** is the molar mass M(Mg).

The value for **m** was given as 25.00 g, and the Molar Mass M(Mg) was established above. Then,

$$25.00 \text{ g} = n * 24.31 \frac{g}{mol}, \text{ next solve for n:}$$

$$n = \frac{25.00 \text{ } g}{24.31 \frac{g}{mol}}$$

Dealing with units simplifying them is identical to the way numbers or variables are simplified in Math. This will also help with error corrections. In this case, it is certain that units for **n** should be moles.

$$n = \frac{25.00}{24.31} \left(g \div \frac{mol}{g} \right)$$

$$n = \frac{25.00}{24.31} \left(g * \frac{mol}{g} \right)$$

$$n = \frac{25.00}{24.31} \quad (g * \frac{mol}{g})$$

The answer is n = 1.028 mol.

Solution 7

The chemical formula for glucose was given in the question: $C_6H_{12}O_6$.

1 1A								18 8A
1 **H** 1.008	2 2A		13 3A	14 4A	15 5A	16 6A	17 7A	2 **He** 4.003
3 **Li** 6.941	4 **Be** 9.012	•••	5 **B** 10.81	6 **C** 12.01	7 **N** 14.01	8 **O** 16.00	9 **F** 19.00	10 **Ne** 20.18
11 **Na**	12 **Mg**		13 **Al**	14 **Si**	15 **P**	16 **S**	17 **Cl**	18 **Ar**

Molar Masses of individual atoms are read from the periodic table and plugged into the calculation.

$M(C_6H_{12}O_6) \quad = M(C_6) + M(H_{12}) + M(O_6)$

$= 6M(C) + 12M(H) + 6M(O)$

$M(C) = 12.01$, $M(H) = 1.008$, $M(O) = 16.00$

Thus,

$M(C_6H_{12}O_6) \quad = 6M(C) + 12M(H) + 6M(O)$

$= 6*12.01 + 12*1.008 + 6*16.00$

$= 72.06 + 12.10 + 96.00,$

and after rounding:

$$= 180.16$$

The answer is $M(C_6H_{12}O_6) = 180.2 \dfrac{g}{mol}$.

Solution 8

The chemical formula for Manganese(IV) Sulfate is $Mn(SO_4)_2$.

The (SO_4) group appears twice. Think of this chemical formula as $Mn + (SO_4) + (SO_4)$.

Molar Masses of individual atoms are read from the periodic table and plugged back into the calculation.

						5	6	7	8
						B	C	N	O
						10.81	12.01	14.01	16.00
						13	14	15	16
7	8	9	10	11	12	Al	Si	P	S
						26.98	28.09	30.97	32.07
25	26	27	28	29	30	31	32	33	34
Mn	Fe	Co	Ni	Cu	Zn	Ga	Ge	As	Se
54.94	55.85	58.93	58.69	63.55	65.38	69.72	72.59	74.92	78.96
43	44	45	46	47	48	49	50	51	52

$M(Mn) = 54.94$, $M(S) = 32.07$, $M(O) = 16.00$

Thus,

$$
\begin{aligned}
M(Mn(SO_4)_2) &= M(Mn) + 2M(SO_4) \\
&= M(Mn) + 2[M(S) + 4M(O)] \\
&= 54.94 + 2(32.07 + 4*16.00) \\
&= 54.94 + 2*96.07 = 54.94 + 192.1 \\
&= 247.0
\end{aligned}
$$

The answer is $M(Mn(SO_4)_2) = 247.0 \frac{g}{mol}$.

Solution 9

The molar mass of SiO_2 can be calculated from the molar masses of the individual atoms in the chemical formula.

Think of this chemical formula as

$Si + O_2 = Si + O + O$.

5	6	7	8	9
B	C	N	O	F
10.81	12.01	14.01	16.00	19.00
13	14	15	16	17
Al	Si	P	S	Cl
26.98	28.09	30.97	32.07	35.45

$M(Si) = 28.09$, $M(O) = 16.00$

$$M(SiO_2) = M(Si) + M(O_2)$$
$$= M(Si) + 2M(O)$$
$$= 28.09 + 2*16.00 = 28.09 + 32.00$$
$$= 60.09$$

$M = 60.09 \frac{g}{mol}$.

The mass can then be extracted from the formula

m = n * M, see *(Equation 11)*,

and since this substance was given in one mole: n = 1 mol. The values can be plugged into the m = n * M formula to calculate m:

$$m = 1 \text{ mol} * 60.09 \frac{g}{mol}$$

$m = 1 * 60.09 \text{ mol} * \dfrac{g}{mol}$, the mol units are simplified,

and the answer is:

$m = 60.09$ g of SiO_2 are present in 1 mole.

Solution 10

Don't let the wording "moles of this molecule" confuse you. The molecules we are examining in this context are O_2 molecules.

The molar mass of O_2 can be calculated from the molar masses of the individual atoms in this diatomic[10] molecule. Think of O_2 as O + O.

5	6	7	8	9	10
B	C	N	O	F	Ne
10.81	12.01	14.01	16.00	19.00	20.18
13	14	15	16	17	18
Al	Si	P	S	Cl	Ar

$M(O) = 16.00$

$M(O_2)$	$= M(O_2)$
	$= 2M(O)$
	$= 2*16.00$
	$= 32.00$

$$M = 32.00 \frac{g}{mol}.$$

---

[10] diatomic means containing two atoms.

128

Since the question stated this substance was given in n = 0.0125 moles, seems we have everything necessary to calculate the mass as **m = n * M**, see *(Equation 11)*.

$$m = 0.0125 \text{ mol} * 32.00 \frac{g}{mol}$$

$$m = 0.0125 * 32.00 \text{ } \cancel{mol} * \frac{g}{\cancel{mol}}$$

After simplifying the mol units, the answer is:

m = 0.4000 g of O_2 are present in 0.0125 mol.

Solution 11

Isotope abundances contribute to the Atomic Weight calculation in a weighted type of average.

Review the *Atomic Weight* concept and *Calculating Weighted Averages*. The first two columns were given as part of the question. The third column is extracted from the hyphenated number in the isotope notation.

isotope	abundance	Nucleon count = Atomic Mass
X-28	92.21%	28 amu
X-29	4.70%	29 amu
X-30	3.09%	30 amu

The Atomic Weight is the weighted average of the Atomic Masses:

$$\frac{92.21\% * 28\ amu\ +\ 4.70\% * 28\ amu\ +3.09\% * 30\ amu}{92.21\% \ +\ 4.70\% \ +\ 3.09\%},$$

Atomic Weight $= \dfrac{2806.18\ \% * amu}{100.00\ \%}$, and after simplifying the percent units we get:

Atomic Weight = 28.0618 amu.

After rounding we are left with the final answer:

Atomic Weight = 28.5 amu (or u).

Solution 12

The way to solve this problem is to look up this element in the periodic table by trying to spot not the Atomic Number as usual, but rather the Atomic Weight calculated in the previous question: 28.5 amu.

After scanning the periodic table decimal values (bottom of each cell), it is noticeable that Silicon's Atomic Weight 28.09 amu comes very close.

5	6	7
B	C	N
10.81	12.01	14.01
13	14	15
Al	Si	P
26.98	28.09	30.97
31	32	33
Ga	Ge	As

There is no other element that is quite so close in value to the Atomic Weight of Silicon.

Thus, the answer is element X = Silicon.

Solution 13

^{30}Si is an isotope of Silicon.

After looking up the Silicon atom in the periodic table it is found to have an **Atomic Number** of 14.

5	6	7
B	C	N
10.81	12.01	14.01
13	14	15
Al	Si	P
26.98	28.09	30.97
31	32	33
Ga	Ge	As

Review the section on *Isotope Notation* to see that for any atom X, the isotope notation reflects:

$$_Z^A X,$$

A = the full nucleon count;

Z is the p$^+$ count, or atomic number.

The full mass of the atom is

A = Z + number of neutrons, based on *(Equation 2)*

and if we treat this as an equation and solve for the number of neutrons, we get:

number of n^0 = A – Z.

The isotope given was

$$_{14}^{30} Si,$$

A = 30 the nucleon count;

Z = 14 is the p$^+$ count = or atomic number.

Thus,

131

number of $n^0 = A - Z = 30 - 14 = 16$

The answer is: the ^{30}Si isotope has 14 p^+ and 16 n^0.

Solution 14

Part (a)

Since 100.00 g of sulfuric acid were given, the formula

$$m = n * M, \text{ see } \textit{(Equation 11)}$$

can be used to calculate the moles **n**. The value for **m** is given as 100.00 grams, and **M** can be computed for $M(H_2SO_4)$ by using the periodic table for each atom in the chemical formula. The number of moles is based on:

$$n = \frac{m}{M} \text{ see } \textit{(Equation 10)}$$

Let's start by calculating the Molar Mass for the sulfuric acid: H_2SO_4. The atoms in this formula were identified in the table cells below.

1A													8A
1								13	14	15	16	17	2
H	2							3A	4A	5A	6A	7A	**He**
1.008	2A												4.003
3	4					5	6	7	8	9	10
Li	**Be**	•••				**B**	**C**	**N**	**O**	**F**	**Ne**
6.941	9.012					10.81	12.01	14.01	16.00	19.00	20.18
11	12					13	14	15	16	17	18
Na	**Mg**					**Al**	**Si**	**P**	**S**	**Cl**	**Ar**
22.99	24.31					26.98	28.09	30.97	32.07	35.45	39.95
19	20					31	32	33	34	35	36
K	**Ca**					**Ga**	**Ge**	**As**	**Se**	**Br**	**Kr**

$M(H) = 1.008$, $M(S) = 32.07$, $M(O) = 16.00$.

Thus,

$$M(H_2SO_4) = M(H_2) + M(S) + M(O_4)$$
$$= 2M(H) + M(S) + 4M(O)$$
$$= 2*1.008 + 32.07 + 4*16.00$$
$$= 2.016 + 32.07 + 64.00$$
$$= 98.09$$

Review the _Molar Mass Definition_ to remember the units for Molar Mass are $\frac{g}{mol}$.

$M(H_2SO_4) = 98.09 \frac{g}{mol}$, and gets plugged into the formula:

$$n = \frac{m}{M} = \frac{100.00 \ g}{98.09 \ \frac{g}{mol}}$$

Units must be verified, they are expected to yield moles.

$$n = \frac{100.00 \ \cancel{g}}{98.09 \ \frac{\cancel{g}}{mol}}$$

$$n = 1.019 \ \frac{1}{\frac{1}{mol}} = 1.019 \ (1 * \frac{mol}{1})$$

$$n = 1.019 \ mol$$

The answer is 100.00 g of sulfuric acid H_2SO_4 contain 1.019 moles of molecules.

Part (b)

To count the total number of atoms in these molecules one must first know how many atoms are present in one H_2SO_4 molecule. In this formula there are:

- 2 Hydrogens,
- 1 Sulfur, and
- 4 Oxygens.

In total, there are 2 + 1 + 4 = 7 atoms.

Thus, in the 1.019 moles found in Part (a) there are

1.019 * 7 moles of atoms.

The answer is thus 7.133 moles of various atoms that make the molecule.

Solution 15

Part (a)

Since the chemical in question, CO_2 is a molecule, it makes sense to stay within that context if not stated otherwise. Thus, the moles must relate to molecules rather than individual atoms.

The mass of CO_2 is given as **m** = 500.00 g, and if the Molar Mass of one molecule **M(CO2)** was known, then the number of moles **n** can be calculated by:

$$n = \frac{m}{M\,(CO_2)} \text{ moles, based on } \textit{(Equation 10)}.$$

For this to work, the Molar Mass of CO_2 must be calculated. Using the periodic table cells of the individual atoms helps.

5	6	7	8	9
B	C	N	O	F
10.81	12.01	14.01	16.00	19.00
13	14	15	16	17
Al	Si	P	S	Cl

$M(C) = 12.01$, $M(O) = 16.00$

$$M(CO_2) = M(C) + M(O_2)$$
$$= M(C) + 2M(O)$$
$$= 12.01 + 2*16.00$$
$$= 44.01$$

Next, plug in all values found so far.

$$n = \frac{m}{M\ (CO_2)}\ \text{moles}$$

$$n = \frac{500.00\ g}{44.01\ \frac{g}{mol}}$$

$$n = \frac{500.00}{44.01} \left(g \div \frac{g}{mol}\right)$$

$$n = \frac{500.00}{44.01} \left(g * \frac{mol}{g}\right)$$

$$n = \frac{500.00}{44.01} \left(\cancel{g} * \frac{mol}{\cancel{g}}\right)$$

Therefore, the answer is **n** = 11.36 mol.

Part (b)

To find the number of molecules, let's call it **r**, found in the given amount 500 g, the moles found in Part (a) are multiplied by Avogadro's Number, see *(Equation 6)*.

$r = n * N_A$

$r = 11.36 \; mol * 6.022 \times 10^{23} \; \dfrac{molecules}{mol}$

$r = 11.36 * 6.022 \times 10^{23} \; mol * \dfrac{molecules}{mol}$

$r = 11.36 * 6.022 \times 10^{23} \; mol * \dfrac{molecules}{mol}$

The answer is 500.00 g of CO_2 contain:

$r = 68.410 \times 10^{23}$ molecules.

Solution 16

This problem can be solved based on the Molar Mass, mass in grams, and mole formula:

m = n * M, see *(Equation 11)*.

we know that:

m = 15.02 g, and

n = 0.0915 mol.

As for the molar mass in the formula we know that

136

$$M = M(X(NO_3)_2).$$

This is then based on three elements: X, N, and O. Two of these, Nitrogen and Oxygen can be looked up in the periodic table.

5	6	7	8	9
B	C	N	O	F
10.81	12.01	14.01	16.00	19.00
13	14	15	16	17
Al	Si	P	S	Cl

$M(N) = 14.01$, $M(O) = 16.00$

$$
\begin{aligned}
M(X(NO_3)_2) &= M(X) + M((NO_3)_2) \\
&= M(X) + 2M(NO_3) \\
&= M(X) + 2[M(N) + M(O_3)] \\
&= M(X) + 2[M(N) + 3M(O)] \\
&= M(X) + 2M(N) + 2*3M(O) \\
&= M(X) + 2*14.01 + 6*16.00 \\
&= M(X) + 28.02 + 96.00 \\
&= M(X) + 124.02
\end{aligned}
$$

After plugging all the values that we know so far into the formula, we get:

$m = n * M$

$15.02 \text{ g} = 0.0915 \text{ mol} * M(X(NO_3)_2) \dfrac{g}{mol}$

$15.02 \text{ g} = 0.0915 * (M(X) + 124.02) \ mol * \dfrac{g}{mol}$

$$15.02 \text{ g} = 0.0915 * (M(X) + 124.02) \; mol * \frac{g}{mol}$$

Divide both sides of the equation by 0.0915.

$$\frac{15.02}{0.0915} \text{ g} = (M(X) + 124.02) \text{ g}$$

$$164.2 = M(X) + 124.02$$

Subtract 124.02 from both sides of the equation.

$$164.2 - 124.02 = M(X)$$

$$M(X) = 40.2$$

Therefore, the mystery element X's molar mass equals $40.2 \frac{g}{mol}$.

Next, a search in the periodic table must examine how many elements have a Molar Mass of approximately the found value: $40.2 \frac{g}{mol}$. However, the periodic table doesn't list Molar Mass values but rather the Atomic Weights of the atoms. This is not a problem since the two values are identical, the only difference between the them is the unit.

Instead of $40.2 \frac{g}{mol}$ one must look for 40.2 amu.

To find this atomic weight, the periodic table can either be scanned visually like in this example, or a digital search can be performed on a webpage such as Encyclopedia Britannica's at:

https://www.britannica.com/science/atomic-weight,

In this example all periodic table cells with atomic weights close to 40.2 amu were colored (see the figure on the next page).

There is no point to look passed the fourth row of the periodic table because atomic weights from there on are always larger than 85 amu.

1	2 2A	3	4	5	6	7	8	9	10	11	12	13 3A	14 4A	15 5A	16 6A	17 7A	18
1 H 1.008																	2 He 4.003
3 Li 6.941	4 Be 9.012											5 B 10.81	6 C 12.01	7 N 14.01	8 O 16.00	9 F 19.00	10 Ne 20.18
11 Na 22.99	12 Mg 24.31											13 Al 26.98	14 Si 28.09	15 P 30.97	16 S 32.07	17 Cl 35.45	18 Ar 39.95
19 K 39.10	20 Ca 40.08	21 Sc 44.96	22 Ti 47.88	23 V 50.94	24 Cr 52.00	25 Mn 54.94	26 Fe 55.85	27 Co 58.93	28 Ni 58.69	29 Cu 63.55	30 Zn 65.38	31 Ga 69.72	32 Ge 72.59	33 As 74.92	34 Se 78.96	35 Br 79.90	36 Kr 83.80
37 Rb 85.47	38 Sr 87.62	39 Y 88.91	40 Zr 91.22	41 Nb 92.91	42 Mo 95.94	43 Tc (98)	44 Ru 101.1	45 Rh 102.9	46 Pd 106.4	47 Ag 107.9	48 Cd 112.4	49 In 114.8	50 Sn 118.7	51 Sb 121.8	52 Te 127.6	53 I 126.9	54 Xe 131.3

140

The colored cells and their elements were listed in the next table. Their atomic weights appear in the second column as read from the periodic table. The atomic weight of element X was found to be 40.2 amu.

The third column calculates how much each of the colored elements differs from the mystery element X.

Element	Atomic Weight in amu	The Difference: 40.2 − atomic weight
Ar	39.95	$40.2 - 39.95 = 0.3$
K	39.10	$40.2 - 39.10 = 1.1$
Ca	40.08	$40.2 - 40.08 = 0.1$

Another wording for this is how much each element deviates from X.

After examining the differences, it becomes clear that the smallest difference, or rather the value closest to that of X appear in the row for **Calcium**, with an atomic weight of 40.08 amu.

There is no other element in the periodic table with a closer value to 40.08, thus the mystery element is **Ca**.

Solution 17
Part (a)

The atoms participating in the fructose molecule $C_6H_{12}O_6$ were colored in the periodic table below.

1								18
1A								8A
1	2	13	14	15	16	17		2
H	2A	3A	4A	5A	6A	7A		He
1.008								4.003
3	4	5	6	7	8	9		10
Li	Be	B	C	N	O	F		Ne
6.941	9.012	10.81	12.01	14.01	16.00	19.00		20.18
11	12	13	14	15	16	17		18
Na	Mg	Al	Si	P	S	Cl		Ar

$M(H) = 1.008$, $M(C) = 12.01$, $M(O) = 16.00$

Thus,

$$
\begin{aligned}
M(C_6H_{12}O_6) &= M(C_6) + M(H_{12}) + M(O_6) \\
&= 6*M(C) + 12*M(H) + 6*M(O) \\
&= 6*12.01 + 12 * 1.008 + 6*16.00 \\
&= 72.06 + 12.096 + 96.00 \\
&= 180.16
\end{aligned}
$$

The answer is $M(C_6H_{12}O_6) = 180.16 \dfrac{g}{mol}$.

Part (b)

The molar mass of the mystery molecule $(CH_2O)_x$ is $M((CH_2O)_x) = x * M(CH_2O)$.

To find $M(CH_2O)$ before the repetitions are applied, we can either divide the result in Part (a) by 6, because $C_6H_{12}O_6 = (CH_2O)_6$, or we can recalculate it from scratch as shown.

$$M(CH_2O) \qquad = M(C) + M(H_2) + M(O)$$

142

$$= M(C) + 2*M(H) + M(O)$$

$$= 12.01 + 2 * 1.008 + 16.00$$

$$= 30.03$$

Therefore $M(CH_2O) = 30.03 \ \frac{g}{mol}$, in which case:

$M((CH_2O)_x) \ \frac{g}{mol} = x * M(CH_2O)$ is the same as

$M((CH_2O)_x) \ \frac{g}{mol} = x * 30.03 \ \frac{g}{mol}$

To get $60.06 \ \frac{g}{mol}$ of $M((CH_2O)_x)$, we are looking for an x that satisfies the following equation:

$60.06 \ \frac{g}{mol} = x * 30.03 \ \frac{g}{mol}$

Divide both sides of this equation by the units $\frac{g}{mol}$.

$60.06 = x * 30.03$

Divide both sides of the equation by 30.03.

$\frac{60.06}{30.03} = x$

Thus x = 2.00 repetitions.

Part (c)

The number of repetitions found in Part (b) is x = 2.

Therefore the molecule $(CH_2O)_x$ is the same as $(CH_2O)_2$. This formula can be represented as either $(CH_2O)_2$ or $C_2H_4O_2$. A quick online search reveals the name of this molecule is vinegar, otherwise known in chemistry as acetic acid.

Math Review: Exponents

The number 100 trillion is equivalent to 10 multiplied by itself 14 times, it has a total of 14 zeroes: 100,000,000,000,000 atoms.

Replacing the long number with an exponential notation is more efficient:

$$10^{14}$$

10 is called the **base**,

14 is called an **exponent**, and it is written in a slightly smaller font at the right top corner of the base called a superscript.

The presence of an exponent typically indicates the base is being multiplied by itself repeatedly, unless the base is 0 or 1:

10 x 10 x ... x 10

14 times

The exponent shows the number of **repetitions** in which the base appears in the **product**. The word "product" when used in math means "multiplication".

The base can be any number. Generalized exponent rules use letters instead of numbers. For example, when

145

demonstrating how squaring works instead of stating $3*3 = 3^2$, a more generalized view is described by using a letter instead of the number 3: $b*b = b^2$, indicating the statement is true for all numbers.

When pretending a letter is essentially a number, we call the letter a **pronumeral**. **Pronumerals** are used very frequently in math and science. Calculations in chemistry require one to be well accustomed to pronumeral notation.

In the next example, the pronumeral b represents any number used as the base, and m represents the exponent in the expression b^m, read as "the number b to the power of m".

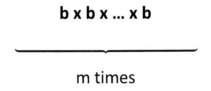

$$b \times b \times \dots \times b$$

m times

Here are a few examples of **b** and **m** combinations that make b^m.

$$b = 3$$
$$m = 2$$
$$\left.\vphantom{\begin{matrix}a\\b\end{matrix}}\right\}\quad 3^2 = 3*3 = 9$$

$$b = 5$$
$$m = 3$$
$$\left.\vphantom{\begin{matrix}a\\b\end{matrix}}\right\}\quad 5^3 = 5*5*5 = 125$$

$$b = 9$$
$$m = 2$$

$$9^4 = 9*9*9*9 = 6{,}561$$

Make sure you don't confuse the base with the exponent.

Exponent Use

10^{14} is a good example of how exponents help with astronomically large numbers.

Exponents can also be used with extremely small numbers, which have long trails of zeros between a *0.* and *some specific number*. For example,

$$0.00000000000004.$$

Such decimal numbers can be converted to fractions. Different colors were used below to emphasize stages of the magnitudes.

$$0.00000000000004 = \frac{4}{100{,}000{,}000{,}000{,}000}$$

After plugging in the earlier value for 10^{14} to replace the denominator: $100{,}000{,}000{,}000{,}000 = 10^{14}$, we get

$$\frac{4}{100{,}000{,}000{,}000{,}000} = \frac{4}{10^{14}} = \frac{4}{10^{14}} = \frac{4*1}{10^{14}} =$$

$$4 * \frac{1}{10^{14}}$$

The interesting part in this result is:

$$\frac{1}{10^{14}}$$

The form "*1 over a power of ten*", is a frequently reoccurring theme in exponents. It should be noticed whenever possible, even when it's somewhat hidden:

$$\frac{4}{10^{14}} \quad \curvearrowright \quad 4 * \frac{1}{10^{14}}$$

The Mathematical convention is to allow replacement of the "*1 over a power of ten*" part with a more compact representation. The notation gets rid of the 1 at the top and removes the fraction line. For this, **the exponent's sign must be flipped** from positive to negative, and if negative then to positive. In other words:

$$\frac{1}{10^{14}} = 10^{-14} \quad \text{and} \quad \frac{1}{10^{-14}} = 10^{14}$$

or more generally,

$$\frac{1}{10^{x}} = 10^{-x}, \text{whatever the sign of } x \text{ may be.}$$

Replacing the zeros of extreme sized numbers leads to either

- **positive** exponents for very **large** numbers, or
- **negative** exponents for extremely **small** numbers.

In conclusion, the previous example yields:

$$0.00000000000004 =$$

148

$$\frac{4}{100,000,000,000,000} =$$

$$\frac{4}{10^{14}} = 4 * \frac{1}{10^{14}} =$$

$$4 * 10^{-14}$$

There are a few rules that govern exponent calculations. The next section dives in and explores them in detail.

Rules for Exponents with Identical Bases

The Power of Zero

By convention, the power zero reduces any base to the number 1. This is always the case.

$$b^0 = 1, b \neq 0$$

There are several mathematical reasons why the power of zero must be 1, but none of concern in this book's context.

Nevertheless, just for fun we can explore a downward trend of exponents for any base. In the next example we'll use a random number 5, so the specifics can be examined. The number 5 is raised to the power of 3 and a downwards trends follows to predict the value of the next exponent in descending order.

$5^3 = 125,$ $125 \div 5 = 25$ transitions to the next power: 2

$5^2 = 25,$ $25 \div 5 = 5$ transitions to the next power: 1

$5^1 = 5,$ $5 \div 5 = 1$ transitions to the next power: 0

$5^0 = 1$ Notice the **base** is divided by, thus must be **non-zero**.

In conclusion, all numbers raised to the power 0 result in the number 1.

The Power of One

The power one does nothing. Any number can be seen as a multiplication of itself that makes an appearance only once in the product, with no repetitions. The power of one is useful when a number is to be treated as an exponent while using the "*multiplying identical bases*" rule, listed later in this chapter.

$$b^1 = b$$

The power of one should not be confused with the power of zero.

A Power in a Denominator

Any time 1 is divided by a base b that is raised to a power m, the fraction line can be eliminated if the sign of the exponent is flipped.

$$\frac{1}{b^m} = b^{-m}$$

This rule is a convention, it does not need to be understood. Although the minus can be remembered as the opposite direction to which positive exponents grow. For now, memorize the rule as simply a more compact notation that helps avoid fraction lines.

Multiplying Identical Bases

$$b^m * b^n = b^{m+n}$$

The equality is explained below.

b^m represents: b^n represents:

$b * b * \ldots * b$ $b * b * \ldots * b$

_____   _____

m times   n times

Insert the repetitions into the original product to replace the two: b^m and b^n:

$b * b * \ldots * b * b * \ldots * b$

_____

Count all b occurrences.

Together there are m + n bases. The exponent is the total number of repetitions. Thus, the expression equals:

$$b^{m+n}.$$

A Power Raised to Another Power

$$(b^m)^n = b^{m*n}$$

The equality is explained below.

b^m represents: $(...)^n$ represents:

$b * ... * b$ $(...) * ... * (...)$

───────⌣───────   ───────⌣───────

m times   n times

Replace the contents of the brackets with instances of b^m expanded.

$$(b * ... * b) * ... * (b * ... * b)$$

───────────⌣───────────

The brackets repeat n times.

The overall count of b-s in this arrangement is: m red bases, then another m red bases, and so on, repeating n times. Thus, there are m * n repetitions of b. Placing the count into the exponent position yields the result:
$$b^{m*n}.$$

Powers Distribute into Brackets

$$(a * b)^m = a^m * b^m$$

The equality is explained below.

$(...)^m$ represents m repetitions of whatever is in the brackets:

$$(\dots) * \dots * (\dots)$$

$$\underbrace{}$$

m times

The brackets must be replaced by the initial content:

$$ab$$

the result is:

$$(ab) * \dots * (ab)$$

$$\underbrace{}$$

m times

We can remove the brackets and swap the position of multiplied elements, similarly to 2*3 = 3*2. All a -s are moved to the front and b -s to the back:

$$a * a * \dots \quad * \quad b * b * \dots$$

$$\underbrace{} \quad \underbrace{}$$

m times   m times

After replacing the repetitions with their exponent counterparts, the result is $a^m * b^m$.

The method for $a * b$ can be adapted for $a \div b$ almost identically. Here too the power distributes equally into the brackets.

$(a \div b)^m = a^m \div b^m$, or rather

154

$$\left(\frac{a}{b}\right)^m = \frac{a}{b} * \dots * \frac{a}{b} = \frac{a * \dots * a}{b * \dots * b}, \text{ thus}$$

$$\left(\frac{a}{b}\right)^m = \frac{a^m}{b^m}$$

A Power Divided by Another Power

$$\frac{b^m}{b^n} = b^{m-n}$$

The equality is explained below.

b^m represents: b^n represents:

$b * \dots * b$ $b * \dots * b$

m times   n times

$$\frac{b * \dots * b}{b * \dots * b}$$

For this example, we'll assume m > n, though m >= n can be shown to work in a similar way.

The numerator showed m occurrences initially. From those we separate n of them into their own set:

$$m = m - n + n = (m - n) + n$$

These form two groups of instances: n and (m − n).

155

Together these are still repeated m times:

$$n \text{ times} \quad (m - n) \text{ times}$$

$$\frac{b * \ldots * b * b * \ldots * b}{b * \ldots * b}$$

The n bases in the denominator remain unchanged.

Next, there is a group of n pieces in the numerator and another in the denominator, and they simplify.

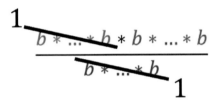

There are $(m - n)$ bases left in the numerator, and the denominator is equal to 1. The result is thus:

$$b * \ldots * b$$

$$m - n \text{ times}$$

After placing the count into the exponent position, the result is:

$$b^{m-n}.$$

All exponent rules should be memorized.

Square roots and other radicals are not covered in this volume.

Math Review: The Rule of Three

The *Rule of Three* takes three known values to calculate a forth. The four values must be arranged as two proportional pairs for it to work. Let's clarify through an example.

Example 1

Imagine one of your exams was graded. There are four values that relate to your grade and are shown in red below.

100%, the grade given for answering everything correctly.

Your grade as a percent (%).

N, the total number of questions present in your exam.

k, the number of questions you have answered correctly.

The four values relate to each other in pairs based on logic, proportions, and the wording of the problem.

Option 1:

	Your grade % is to	**k**, questions answered correctly,
like	**100 %** is to	**N**, the total number of questions.

Option 2

Alternately you could say:

	Your grade % is to	100 %,
like	**k** is to	N, the total number of questions.

Notice that related, or rather proportional elements appear in rows and columns in both options. The easiest way to place the values correctly is to follow the idea that one pair relates **like** to the second pair, based on common sense.

Whenever you say,

	thing 1 is to	thing 2
like	thing A is to	thing B

you are relating a **proportion** between two pairs:

(thing 1, thing 2) is like (thing A, thing B) over rows, and

(thing1, thing A) is like (thing 2, thing B) over columns.

Options 1 and 2 both form a grid with rows and columns. Each of the 4 numbers is positioned in one cell according to the logic described. Let's examine the logic used:

- 100 % corresponds to the N questions because they both represent the total number of questions.
- Your grade % corresponds to the number of questions k you have answered correctly, because they both represent same thing: how many you have answered correctly.
- k is a proportion of the overall number of questions N in the same way that your grade % is a proportion of 100 %.

Values should not be placed randomly in the grid. For instance, k is NOT to 100 % like N is to your grade %. Pairing these over columns or rows would be incorrect.

Typically, problems that require the *Rule of Three* to be solved, provide 3 of the 4 values, and the fourth is considered the unknown and is computable.

In this example we'll say the grade is the unknown and must be calculated. Below is a demonstration of how to calculate by using the grid in *Option 1*.

	x = Your grade % is to	k, questions answered correctly,
like	100 % is to	N, the total number of questions.

A simpler version of the grid is a box, or rather rectangle, with values in every corner.

Step 1

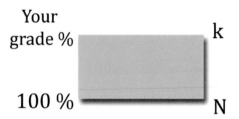

Step 2

Circle the value to be calculated, the unknown. The result is going to be a fraction. Identify the number in the opposite corner as the **denominator** of that fraction. Provide a skeleton of the fraction with what you know so far.

The skeleton of the fraction so far is:

$$\text{Your grade \%} = \frac{?}{N}$$

Step 3

The other unused values are on the diagonal. Multiply them and place their product as the fraction's numerator.

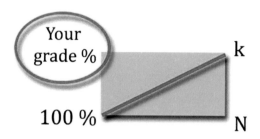

The fraction then becomes:

$$\text{Your grade \%} = \frac{100 * k}{N} \%.$$

This solves the problem because the unknown is isolated on the left, and all the known values are on the right.

Homework: Use Option2 to solve this problem again, then compare results to convince yourself that identical results have been obtained.

Let's explore one more example to solidify the concept.

Example 2

Imagine you and your friends planted 28 flowers together. If you planted 7 flowers, what percent of the total work was completed by you?

Again, you must create a grid with pairing values. Pairing is logical and happens over both rows and columns.

	28 flowers are to	100 %,
like	7 flowers are to	x = your work percent.

Stating the following is also logical.

	28 flowers are to	7 flowers,
like	100% is to	x = your work percent.

Selecting either option will result in identical calculations. The diagram below demonstrates the calculation for the second option after copying the grid values into the corners of a rectangle.

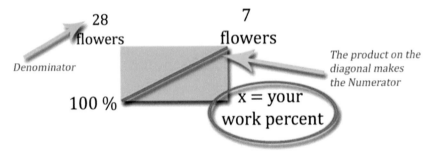

The fraction then becomes:

$$x, \text{your work } \% = \frac{100 * 7}{28}\%$$

After carrying out the multiplication and division, turns out x, your work % equals **25%**.

Does Any Arrangement Work?

The previous two options were both examples of logical layouts. Let's show an option that is **not** logical and should **not** be used but is a common mistake.

Pairing the values as follows is **incorrect**:

162

28 flowers to	7 flowers
x = your work percent to	100%

The total 28 flowers do not relate to 7 flowers as your percent relates to 100% of the work, but rather the opposite: as 100% is to 7 flowers. The total number of flowers is the only value that is like 100%.

What's the Correction?

One possible fix would be to swap 100% with 7 flowers.

28 flowers to		100%
x = your work percent to	after swapping	7 flowers

Another way to think about how to prevent mistakes is to make sure that equivalent values like the

- total number flowers, and
- the 100% they represent

are never arranged in opposite corners of the grid.

Formula Sheets

Exponent Laws

$$b^0 = 1, b \neq 0 \qquad b^m * b^n = b^{m+n}$$

$$b^1 = b \qquad\qquad (b^m)^n = b^{m*n}$$

$$\frac{1}{b^m} = b^{-m} \qquad (a*b)^m = a^m * b^m$$

$$\qquad\qquad\qquad \left(\frac{a}{b}\right)^m = \frac{a^m}{b^m}$$

$$\frac{b^m}{b^n} = b^{m-n}$$

Whenever a base is used in the denominator, it must be non-zero.

Terms, Quick Reference

(listed in logical order, not alphabetical)

Nucleon

A nucleon is either a proton or a neutron.

Nucleon Mass

Nucleon mass Is the average mass of one proton's and one neutron's mass.

Ions

An atom with p^+ count $\neq$ e^- count. Every ion is a charged version of the atom, but not every atom is an ion. Whenever p^+ count $=$ e^- count, the atom is not an ion.

Ion Notation

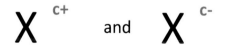

$$X^{c+} \quad \text{and} \quad X^{c-}$$

c+ and c- represent possible charges of the ion, and **X** is the atom's chemical symbol. When c is 1, the sign representing the charge is enough to represent the ion and c is dropped. When c > 1, listing c alongside the sign is required.

Isotope

An atom with p^+ count $\neq$ or $=$ n^0 count. The isotope is a weight variation of neutrons belonging to a specific element. Every atom is an isotope.

Isotope Notation

$$_{Z}^{A}X \quad \text{or} \quad {}^{A}X$$

A is the Mass Number, Z is the p^+ count, and **X** is the atom's chemical symbol. The short version of the notation is shown on the right.

Atomic Number, or

Proton Number,

Symbol Z

The Atomic Number $= p^+$ count and identifies the element uniquely.

Mass Number, or

Atomic Mass, or

Atomic Mass Number, or

Nucleon Number, or

Symbol A

The atomic mass = p^+ count + n^0 count, it relates to an individual isotope. Atomic mass is measured in **amu**. The value can also be found in the top left corner of all isotope notations.

Atomic Weight,

Average Atomic Weight, or

Average Atomic Mass, or

Standard Atomic Weight, or

Symbol $Ar_{standard}$

Atomic Weight is the average of ($p^+ + n^0$) measured in amu units, of isotopes in a sample and relates to the collection of all variations of an atom that are assumed to exist on Earth. It is considered to be a global average rather than relating to a specific sample.

Relative Atomic Mass, or

Symbol Ar

The relative atomic mass is the same as Atomic Weight but less general, applies to specific samples rather than an overall global prevalence.

Mole

The mole is the name given to a specific count used in chemistry. It is logically similar to other famous counts such as: pair (2), dozen (12), baker's dozen (13), myriad (10,000) and googol (10^{100}).

1 mole = 6.022 x 1023 pieces.

Molar Mass of an Element

The Molar Mass of an element is the weight measured in grams for a count of 1 mole of atoms. The result has its own special unit of grams per mole or $\dfrac{g}{mol}$. The value for each atom is read from the periodic table, where it is listed at the bottom of every cell as a decimal number to which the unit $\dfrac{g}{mol}$ is simply assigned as is with no conversions.

Molar Mass of a Chemical Formula

The Molar Mass of a chemical formula is the sum of the Molar Masses of all individual elements participating in the formula, used as many times as elements appear in it. For example, the chemical formula for sucrose, known as table sugar is $C_{12}H_{22}O_{11}$, thus the molar mass of this formula is the total of 12 Carbon molar masses, 22 Hydrogen molar masses, and 11 Oxygen molar masses. The result has the unit of grams per mole or $\frac{g}{mol}$.

Molar Mass vs. grams and moles

The relationship between mass, moles, and the Molar Mass of a sample, is defined by the equation **m = n* M**.

Chemical Constants and Formulas

Nucleon to Electron Mass Ratio

The mass of one nucleon is the same as that of 2,000 electrons.

Proton Mass

The mass of a Hydrogen's proton is:

1.00727647 amu.

Neutron Mass

The mass of a Deuterium, or Hydrogen-2 (^{2}H) neutron is:

1.00866490 amu.

The Atomic Mass Unit to Grams Conversion

$$1 \text{ amu} = 1.6605 \times 10^{-24} \text{ g}$$

Avogadro's Number

Avogadro's number is the count of units in one mole. The notation N_A is used to denote it.

$$1 \text{ mole} = 6.022 \times 10^{23} \text{ units of anything} = N_A$$

Avogadro and AMU to Grams Reciprocal

$$1 \text{ mole} = 6.0221 \times 10^{23} \text{ units} = N_A =$$

$$\frac{1}{1.6605 \times 10^{-24}} =$$

$$\frac{1}{1 \text{ amu to grams conversion}}$$

Molar Mass, Moles and Mass Relation

m = n * M, where

m   the mass of a substance, given for example in grams,

n   the number of moles the substance is given in,

M   if the Molar Mass of the element or chemical formula defining the substance.

Table of Equations

Table of Figures

Index

Errata and Feedback

This book's errata can be found at:

http://sharpseries.ca/chem/errata1.html

Comments and suggestions for future editions are welcome.

74630948R00104

Made in the USA
Middletown, DE
28 May 2018